INTERNET ACTIVITIES FOR GENERAL BIOLOGY

FOR

**BIOLOGY: THE UNITY AND DIVERSITY OF LIFE
BY STARR & TAGGART**

AND

**BIOLOGY: CONCEPTS AND APPLICATIONS
BY STARR**

AND

**HUMAN BIOLOGY
BY STARR AND McMILLAN**

KIM R. FINER
KENT STATE UNIVERSITY, STARK

Wadsworth Publishing Company
I(T)P® An International Thomson Publishing Company

Belmont, CA • Albany, NY • Bonn • Boston • Cincinnati • Detroit • Johannesburg • London
Madrid • Melbourne • Mexico City • New York • Paris • Singapore • Tokyo • Toronto • Washington

Biology Editor: Jack Carey
Assistant Editor: Kristin Milotich
Production Editor: Howard Severson
Print Buyer: Stacey Weinberger
Compositor: Jeff Sargent
Cover Design: Gary Head
Printer: Globus

COPYRIGHT © 1998 by Wadsworth Publishing Company
A Division of International Thomson Publishing Inc.
I(T)P® The ITP logo is a registered trademark under license.

Printed in the United States of America
1 2 3 4 5 6 7 8 9 10

For more information, contact Wadsworth Publishing Company, 10 Davis Drive, Belmont, CA 94002, or electronically at http://www.thomson.com/wadsworth.html

International Thomson Publishing Europe
Berkshire House 168-173
High Holborn
London, WC1V 7AA, England

International Thomson Editores
Campos Eliseos 385, Piso 7
Col. Polanco
11560 México D.F. México

Thomas Nelson Australia
102 Dodds Street
South Melbourne 3205
Victoria, Australia

International Thomson Publishing Asia
221 Henderson Road
#05-10 Henderson Building
Singapore 0315

Nelson Canada
1120 Birchmount Road
Scarborough, Ontario
Canada M1K 5G4

International Thomson Publishing Japan
Hirakawacho Kyowa Building, 3F
2-2-1- Hirakawacho
Chiyoda-ku, Tokyo 102, Japan

International Thomson Publishing GmbH
Königswinterer Strasse 418
53227 Bonn, Germany

International Thomson Publishing Southern Africa
Building 18, Constantia Park
240 Old Pretoria Road
Halfway House, 1685 South Africa

All rights reserved. No part of this work covered by the copyright hereon may be reproduced or used in any form or by any means—graphic, electronic, or mechanical, including photocopying, recording, taping, or information storage and retrieval systems—without the written permission of the publisher.

ISBN 0-534-53023-0

CONTENTS

INTRODUCTION

PART ONE
THE CELLULAR BASIS OF LIFE   1

1. Handedness, Brain Lateralization, and the Scientific Method   3
2. The Periodic Table of the Elements   7
3. Monosaccharides in 3-D   13
4. The Size and Shape of Microscopic Organisms and Cells   15
5. Phagocytosis   19
6. Enzyme Graphics   21
7. Photosynthesis and the Winogradsky Column   23
8. Step-by-Step Glycolysis   27

PART TWO
PRINCIPLES OF INHERITANCE   29

9.   Mitosis   31
10. Meiosis Review   35
11. Mendelian Genetics and the Virtual Fly Lab   39
12. Exploring the Human Genome Map   45
13. Ethical, Legal, and Social Implications of the Human Genome Project 51
14. How Much Do We Need to Know? Ethical Cases in Genetics   57
15. Molecular Biology Vocabulary   63
16. DNA Isolation   67
17. Transcription and Translation   71
18. Cancer: A Loss of Cell Cycle Control   73
19. Plant Biotechnology   77

PART THREE
PRINCIPLES OF EVOLUTION   85

20. A Tour Through the Virtual Paleontology Museum   87
21. Mutation   89
22. Jumping Spiders and the Tree of Life   93

PART FOUR
EVOLUTION AND DIVERSITY   97

23.	Views of the Ocean Floor and Plate Tectonics	99
24.	Emerging Infectious Disease	103
25.	Protist Survey	111
26.	The Diversity of Life: The Fungal and Animal Kingdoms	115
27.	Bryophyta	121
28.	The Beast of Bodmin Moor	125
29.	Primates and Human Evolution	129

PART FIVE
PLANT STRUCTURE AND FUNCTION   137

30.	Plant Movement	139
31.	Seed Dispersal	141
32.	Secondary Plant Products and Poisonous Plants	145

PART SIX
ANIMAL STRUCTURE AND FUNCTION   149

33.	Virtual Frog Dissection	151
34.	Human Cross-Sectional Anatomy	153
35.	Scientific Journals	157
36.	Alzheimer's Disease	159
37.	Cow's Eye Dissection	163
38.	Diabetes	167
39.	Skin Cancer	173
40.	Heart Sounds	177
41.	Virtual Immunology Laboratory	181
42.	Asthma	185
43.	The Fast Food Diet and Nutrition	189
44.	Kidney Disease Search	193
45.	The Visible Embryo: Embryonic Development From Week One Through Week Four	195
46.	Sexually Transmitted Diseases	197

PART SEVEN
ECOLOGY AND BEHAVIOR   201

47.	Human Population Trends	203
48.	Predator-Prey Interactions	207
49.	Critical Ecoregions	209
50.	The Ocean Planet	213
51.	Tropical Rainforests: Here Today, Gone Tomorrow	217
52.	Environmental Technical Information Project	225
53.	Butterfly and Bird Migration	229
54.	Searching the Web on Your Own	241

INTRODUCTION

Welcome to the Internet Activities Workbook for *Biology: The Unity and Diversity of Life*, by Starr/Taggart, *Biology: Concepts and Applications*, by Starr, and *Human Biology*, by Starr/McMillan. The internet exercises in this workbook are intended to augment and supplement your classroom lectures and readings in the text.

NOTES TO THE PROFESSOR

Because many biology topics build upon previously learned knowledge, you may find useful exercises in several different parts of this workbook. Therefore, be sure to scan the table of contents for any applicable exercise before settling on one activity that seems to suit your needs. We have tried to include a variety of exercises which in turn require students to learn in a variety of ways. Interactive dissections, surveys, genetic crosses, laboratory experiments, and notice posting are just some of the activities included in this manual. Many of the activities include a tear out worksheet which may be handed in for evaluation.

The activites also address a variety of student skill levels. Some activities are for the introductory student, whereas other activities will require the student to apply previously acquired knowledge in order to successfully complete the activity.

WHAT STUDENTS SHOULD KNOW

In these activities you will be traveling to many different sites located throughout the world. You may occasionally feel that you have arrived at a particular site through a very circuitous route. However, there really is a logic to the path you have taken. On the way to many sites you will pass through additional sites. Although the activity is not specifically written for those sites you pass through, you may notice some interesting information in a "pass-through" site. Perhaps you will return to that site on your own when you have more time. Many of the "pass through" sites may be valuable resources for term papers or classroom discussion.

You should know how to use your browser and world wide web tools. Before you begin any of these activities, you should familiarize yourself with the toolbar and buttons of your web browser, particularly the "Back" and "Forward" commands.

The internet is a very busy place, and it's becoming even busier. Therefore, you should avoid using the Internet during the peak times of the day. Late at night or early in the morning is when you'll get the fastest response from the net. Many times the net will be so busy that you will receive a "time out" message from your browser. This simply means that the host site could not connect to the address you have requested in the appropriate time. Try that same address at a less peak time and your search will probably be successful. Another common message from your browser will be "URL not found". This may mean that you have typed in the address incorrectly. Check your typing, and be sure you have included all correct symbols, and lower or uppercase letters.

Many sites on the internet now have both sound and video files associated with them. If you have the appropriate hardware, a soundcard, and video viewer as well as necessary software, you can configure viewers, or access plug-ins to hear sound or view video at your site. One word of caution, these sound and video files can be very large and take several minutes to load.

HARDWARE AND SOFTWARE REQUIREMENTS

You will need access to a computer with software that will provide you entry on to the World Wide Web. Universities with internet access should have these software programs loaded on the computers which are available to students. Ask your instructor for the location of the computer laboratories and account information. Some universities permit dial up access. This means that students with a computer and a modem at home have the ability to log into the school's computer's from a remote location.

HARDWARE/SOFTWARE DIAL UP FROM HOME.

It is recommended that you have at least a 28.8K baud modem. We have created this workbook using Netscape Navigator 3.0 and Microsoft Internet Explorer as our web browsers. Some of these sites will not look right if accessed using other browsers. To download the latest version of Netscape (the licensing agreement allows for unlimited academic use), you can go to http://home.netscape.com/. Instructions for on-line installation are provided.

WHAT IS THE WORLD WIDE WEB AND HOW DO YOU USE IT?

The World Wide Web is a dynamic resource that contains a wealth of information on almost any topic imaginable. Surfing the net can either be fun or frustrating, but it is always very time consuming. Instead of wondering aimlessly through sites of questionable worth, this activity book attempts to steer you down the most direct path to a particular body of knowledge.

Using the web is simple. Once you are on a computer connected to the internet you connect to a program called a browser. The browser retrieves the document you request. The documents, or pages, will have highlighted words, photos, and graphics that you can select. By selecting these highlighted areas, you will be linked to another document. Each time you click on a link you will be traveling to another location. These locations can be scattered throughout the world. To travel around the web you will be using URLs (universal resource locators) or addresses to connect to specific files on remote computers. URLs look like this: http://machine.domaine/directory/file.html.

If you frequently return to a particular site, you can "bookmark" the site. Bookmarks are stored URLs. You create a bookmark by going to a particular site and then selecting the "bookmark" command on your browser's tool bar. Instead of typing URLs many times over, these addresses can then be accessed by simply selecting the name of the site in your bookmark list.

UPDATES ON THE WADSWORTH PUBLISHING HOME PAGE

Before you begin each activity, it is imperative that you check the Wadsworth Biology Resource Center Home Page for any updates on URLs or hyperlinks. As previously stated, the web is a constantly changing place. Many sites outgrow their homes and are moved to mirror sites at other locations. Some sites may be eliminated, and some hyperlinks may no longer be available. As the web continues to grow, new sites are being created or altered, and we wouldn't want you to miss any of them.

P A R T O N E

THE CELLULAR BASIS OF LIFE

1

HANDEDNESS, BRAIN LATERALIZATION, AND THE SCIENTIFIC METHOD

INTRODUCTION
In humans the cerebrum is the largest portion of the brain. It is divided into right and left hemispheres. The two halves are not exactly alike. Each hemisphere appears to have functional specialization. This functional specialization is referred to as brain lateralization. In 70 to 95 percent of humans, speech and language map to the left hemisphere of the brain; however, in 5 to 30 percent of humans, speech and language do not map to the left side of the brain. Language comprehension is found in a special area of the left hemisphere known as Wernicke's area, named for its discoverer, J. Wernicke. Speaking ability is found in an area originally described by Paul Broca known as Broca's area. In 1940, Paul Broca suggested that an individual's handedness was opposite the specialized hemisphere involved in speech and language.

Determining the hand preference of a subject or patient is important to clinical neurologists because it is still considered a marker for hemispheric dominance for speech and language. The most commonly used methodology for determining handedness is via a questionnaire and self-evaluation.

Research into this area requires a broad survey of a large population. What better way to reach a large, diverse group than through the Internet?

CHECK FOR URL UPDATES
Before beginning the activities in this chapter, check the Wadsworth Biology Resource Center home page for URL updates and information:

http://www.wadsworth.com/biology

Once you have arrived at the Wadsworth Biology Resource Center home page, select General Biology and locate this book. Any updates should appear.

ACTIVITY

In this activity you will take part in a research project that attempts to define handedness. The survey will take from 15 to 20 minutes to complete. The first part of the survey is about hand usage for manual tasks, and the second part of the survey asks for demographic information. As you read through the questions, think about the various aspects of this research and the research approach. How do you think that these questions fulfill the requirements of the scientific method?

This research protocol has been approved by the Indiana University Campus Committee for the Protection of Human Subjects.

PART 1

1. Using Netscape, or another Web browser, type in the following address at the URL prompt:

   http://www.indiana.edu/~primate/brain.html

2. Read the introduction to find out what handedness has to do with brain lateralization.

PART 2

1. Scroll down the page and select Hand Preference Questionnaire.
2. The survey on handedness involves responding to multiple choice questions. It will take 10 to 15 minutes to complete. As you answer the questions, make notes of those that you find unusual or off the subject. Try to think about how these questions fit into the researcher's hypothesis.
3. Once you have completed the survey, answer the questions on the Handedness, Brain Lateralization, and the Scientific Method Worksheet found at the end of this chapter.

PART 3

Your instructor may want to initiate a classroom discussion using student responses concerning the quality and utility of the research as well as the investigators' approach to the problem via the scientific method.

Follow the directions from your instructor regarding the submission of your worksheets. Some instructors will want them e-mailed; others will want them handed in.

HANDEDNESS, BRAIN LATERALIZATION, AND THE SCIENTIFIC MWTHOD WORKSHEET

Name _____ Date _____

Class/Section _____

1. Can you determine the hypothesis of this particular research project?

2. How do the questions from the investigator's survey relate to the hypothesis of the project?

3. Name some additional experiments that you feel would be necessary for the researcher to fulfill his or her hypothesis.

4. How could the results of this survey be used to evaluate brain lateralization?

2

THE PERIODIC TABLE OF THE ELEMENTS

INTRODUCTION
On the surface, a human and a rock seem to have very little in common. Yet at their most basic level, both humans and rocks are composed of chemical elements. There are now 112 recognized elements organized into eight different groups. Twenty-two of these elements are synthetic whereas the remaining ninety elements are found in nature in either a solid, liquid, or gaseous state. In the earth's crust, eight elements account for 98 percent of all matter. Approximately eleven elements account for the majority of chemicals in the human body.

Each element is composed of a specified number of protons, neutrons, and electrons. The elements are numbered according to the number of protons, and then organized by number into the periodic table of the elements.

CHECK FOR URL UPDATES
Before beginning the activities in this chapter, check the Wadsworth Biology Resource Center home page for URL updates and information:

http://www.wadsworth.com/biology

Once you have arrived at the Wadsworth Biology Resource Center home page, select General Biology and locate this book. Any updates should appear.

ACTIVITY
In this activity you will access the WebElements site. This site is maintained by the Department of Chemistry at the University of Sheffield, United Kingdom. You will begin this activity on the periodic table by selecting an element and then accessing the general properties, as well as the biological and geologic facts associated with that particular element.

PART 1

1. Using Netscape, or another Web browser, type in the following address at the URL prompt:

   http://www.shef.ac.uk/chemistry/web-elements

2. Once at this site, select the periodic table icon on the right side of the screen. Select the element Zn (zinc) on the periodic table.

3. For general information, under the heading "Background information" select either Description, Uses, or Historical to answer the questions on the Periodic Table Worksheet.

4. Read about the element and its properties and answer the questions on the Periodic Table Worksheet.

5. Scroll back up to the top of the page and select Biological.

6. Read about the biological facts of the element and then answer the questions on the Periodic Table Worksheet.

PART 2

Repeat the above steps by searching for information on the elements P (phosphate), Xe (xenon), and C (carbon).

Note: If you can configure a viewer for sound, you can listen to the correct pronunciation of the elements.

Follow the directions from your instructor regarding the submission of your worksheets. Some instructors will want them e-mailed; others will want them handed in.

PERIODIC TABLE WORKSHEET

Name _____ Date _____

Class/Section _____

ZN (ZINC)

General Properties
1. List the date and discoverer of this element.

2. What is the standard state of this element?

3. Describe two major uses of this element.

Biological Properties
1. Is this element toxic to humans?

2. Of blood, bone, liver, or muscle, which has the greatest concentration of this element?

3. What is the total mass of this element in an average, 70kg human?

P (PHOSPHATE)

General Properties

1. List the date and discoverer of this element.

2. What is the standard state of this element?

3. Describe two major uses of this element.

Biological Properties

1. Is this element toxic to humans?

2. Of blood, bone, liver, or muscle, which has the greatest concentration of this element?

3. What is the total mass of this element in an average, 70kg human?

PERIODIC TABLE WORKSHEET, CONT.

XE (XENON)
General Properties
1. List the date and discoverer of this element.

2. What is the standard state of this element?

3. Describe two major uses of this element.

Biological Properties
1. Is this element toxic to humans?

2. Of blood, bone, liver, or muscle, which has the greatest concentration of this element?

3. What is the total mass of this element in an average, 70kg human?

C (CARBON)

General Properties

1. List the date and discoverer of this element.

2. What is the standard state of this element?

3. Describe two major uses of this element.

Biological Properties

1. Is this element toxic to humans?

2. Of blood, bone, liver, or muscle, which has the greatest concentration of this element?

3. What is the total mass of this element in an average, 70kg human?

3

MONOSACCHARIDES IN 3-D

INTRODUCTION
Carbohydrates are the most abundant biological compounds. These compounds consist of carbon, hydrogen and oxygen organized in a variety of ways. Cells use carbohydrates as structural building blocks and energy resources. Biochemically, we recognize three different groups of carbohydrates, these include the monosaccharides, or simple sugars, the oligosaccharides, a short chain of two or more sugars and the polysaccharides, consisting of long chains or branches of sugar units. Glucose is an example of a monosaccharide, sucrose an oligosaccharide, and starch is an example of a polysaccharide. Although represented as two dimensional flat units in most texts, the carbohydrates form a group of diverse three dimensional molecules. Because the carbon atoms can rotate freely around their chiral centers, the carbohydrates have a full range of steriometric isomers. This diversity of form allows the carbohydrates to interact in a variety of ways with neighboring atoms or compounds allowing this group to play an important role in many cellular processes.

Monosaccarides can further be divided and classified into various groups depending upon their chemical structures. Glucose is considered a hexose because it contains six carbon atoms. Ribose on the other hand is a pentose with only five carbon atoms. We also recognize biologically important tetroses (four carbons) and trioses (three carbons). Monosaccharides may also be named or classified as either aldoses or ketoses depending upon the functional group attached to the molecule. An aldose (glucose) will contain a functional aldehyde group whereas the ketose (Fructose) contains a ketone group. Monosaccharides can be graphically represented as either the D-form or the L-form, referring to the physical orientation of the atoms in the molecule. In this activity you will access the Monosaccharide Browser. While in the browser you will view space filling projections of these molecules. You can view ketones, aldoses, and a variety of monosaccharide lengths. In addition you can alter the orientation or chirality of the molecule by flipping the image to view both D and L forms. There is no activity worksheet for this exercise. The objective is to browse and view at least 4 different monosaccharides, including both their D and L forms. Concentrate on recognizing the differences in structure, and enjoy the diversity.

CHECK FOR URL UPDATES

Before beginning this activity, check on the Wadsworth Biology Resource Center homepage for URL updates and information:

http://www.wadsworth.com/biology

Once you have arrived at the Wadsworth Biology Resource Center homepage, select <u>General Biology</u> and locate this book. Any updates should appear.

ACTIVITY

1. Using Netscape or another Web browser, type the following address at the URL prompt:

   http://bmbwww.leeds.ac.uk/designs/monosac/home.htm

2. Select <u>alphabetical index</u>.

3. Select <u>Aldoses</u>.

4. Choose a monosaccharide from the list of aldoses.

5. View the aldose and take note of the location of the aldehyde group.

6. Select the <u>Reverse All</u> button to view the L form of the carbohydrate.

7. Using the Back button of your browser return to the index.

8. Select a ketose for viewing.

9. Repeat steps 4-6 for the ketose.

10. Return to the index and select for viewing a tetrose.

11. Is the tetrose you selected a ketose or aldose?

12. Finish the exercise by selecting a triose for viewing. Be sure to note the number of carbons, the type of bonding, and the presence of certain functional groups.

4

THE SIZE AND SHAPE OF MICROSCOPIC ORGANISMS AND CELLS

INTRODUCTION
Organisms and cells come in a wide variety of shapes and sizes, from microscopic geometric viruses measuring 60 to 80nm to single-celled eucaryotic protistians like the paramecium measuring in at approximately 200μm. Although viruses are considered acellular and nonliving, they have been extensively studied and described and are usually included in a discussion of cellular structure and diversity.

CHECK FOR URL UPDATES
Before beginning the activities in this chapter, check the Wadsworth Biology Resource Center home page for URL updates and information:

http://www.wadsworth.com/biology

Once you have arrived at the Wadsworth Biology Resource Center home page, select <u>General Biology</u> and locate this book. Any updates should appear.

ACTIVITY

In this activity you will visit a Web site called *Cells Alive*. This site uses graphics, video, and animation to illustrate cellular structure, size, and activity.

PART 1
1. Using Netscape, or another Web browser, type the following address at the URL prompt:

    http://www.cellsalive.com

2. Scroll down the page until you come to the Contents section of the site. Select <u>How Big Is a...</u> by clicking on the bullet.

3. Compare the sizes of the virus, bacterium, red blood cell, lymphocyte, and human sperm by using the line scale provided.

PART 2
1. After looking at all of the images, go back to the viruses and click on the "Virus" box to enlarge the image.
2. From the enlarged virus graphics screen, hyperlink to the WWW Server for Virology by clicking on the underlined phrase.
3. Once at the virology site, select multimedia library. Select Digitized Electron Micrographs.
4. Select Ebola.
5. View the Ebola, Marburg, and Rabies viruses.
6. On the Size and Shape of Cells and Microorganisms Worksheet at the end of this chapter, describe the physical appearance of these three viruses.

PART 3
If you have the ability to configure a video viewer, the cells alive site has some excellent video of a variety of cell processes. Explore!!

To view more microscopic images of different cell types, visit Nanoworld at the following address:

http://www.uq.oz.au/nanoworld/images_1.html

Follow the directions from your instructor regarding the submission of your worksheets. Some instructors will want them e-mailed; others will want them handed in.

SIZE AND SHAPE OF CELLS AND MICROORGANISMS WORKSHEET

Name _____   Date _____

Class/Section _____

Compare and contrast the size, shape, and structure of the Marburg, Ebola, and Rabies viruses.

5

PHAGOCYTOSIS

INTRODUCTION
The cell membrane is a formidable barrier effectively limiting the passage of molecules into and out of the cell. In order for large substances to pass into a cell, they must enter through a specialized transport process known as endocytosis. Endocytosis allows large substances to enter the cell by going around rather than through the cell membrane.

A well-studied example of endocytosis is the activity of immune cells known as phagocytes. When phagocytic cells ingest bacterial cells, extensions of cytoplasm or pseudopods, surround the target cell and engulf it to form a vesicle. Lytic enzymes are then released into the vesicle destroying the target cell. Debris from the dead cell is then excreted through a process known as exocytosis.

In this activity you will view two movies showing phagocytic cells engulfing bacteria and yeast cells. To view the movies you must have QuickTime™ loaded on your computer. QuickTime™ can be downloaded from the following address:

http://quicktime.apple.com/

CHECK FOR URL UPDATES
Before beginning this activity, check on the Wadsworth Biology Resource Center homepage for URL updates and information:

http://www.wadsworth.com/biology

Once you have arrived at the Wadsworth Biology Resource Center homepage, select General Biology and locate this book. Any updates should appear.

ACTIVITY

1. Using Netscape or another Web browser, type the following address at the URL prompt:

   http://www.cellsalive.com/mac.htm

2. Select <u>Click for movie</u> to view the video (If you are using a modem, this video may take a few minutes to load).

3. To view the second movie, type in the following address at the URL prompt:

   http://www.cellsalive.com/dictyo.htm

4. Select <u>Click for movie</u> to view the video (If you are using a modem, this video may take a few minutes to load).

6

ENZYME GRAPHICS

INTRODUCTION
Enzymes are proteins that function as biological catalysts to quickly drive biochemical reactions to completion yet remain unchanged by the reaction. Enzymes are very selective regarding the compounds (substrates) they interact with. The substrate and the enzyme interact in a very specific manner. This specificity is determined by the three dimensional structure of the enzyme, particularly at the active site. Recall that the monomeric unit of a protein is an amino acid. Chains of amino acids fold and twist forming a three dimensional protein molecule often consisting of helices, coils and pleated sheets. The charged or polar groups of amino acids within the active site allow the enzyme to chemically recognize the substrate.

In this activity you will view three dimensional graphical representations of various enzymes. While you are viewing these graphics, take note of those enzymes which require co-factors (metals or other compounds included in the graphics). Before viewing the enzymes, it is important to review the basic structure of the twenty amino acids.

CHECK FOR URL UPDATES
Before beginning this activity, check on the Wadsworth Biology Resource Center homepage for URL updates and information:

http://www.wadsworth.com/biology

Once you have arrived at the Wadsworth Biology Resource Center homepage, select General Biology and locate this book. Any updates should appear.

ACTIVITY

1. Using Netscape or another Web browser, type the following address at the URL prompt:

   http://www.sirius.com/~johnkyrk/aminoacid.html

2. It will take a few minutes for all of the amino acid graphics to load but once they are loaded each amino acid will appear in sequence. Review the structures as the graphics cycle.

3. Using Netscape or another Web browser, type the following address at the URL prompt:

   http://www.imb-jena.de/IMAGE.html

4. Select <u>Proteins</u>

5. View five different enzymes by selecting the abbreviation for an enzyme. (Note: some of the enzymes have multiple .gif files which can be viewed. Select only one of these files to view. The smallest file will take the least time to load.)

6. As you view the graphics, be sure to note the presence or arrangement of pleated sheets, coils and helixes as well as any co-factors which are present.

7

PHOTOSYNTHESIS AND THE WINOGRADSKY COLUMN

INTRODUCTION

Photosynthesis is the only chemical process in which light energy from the sun can be harvested to produce ATP. When most people think of photosynthesis, the activity of plants come to mind. However, certain bacteria and some protists also have the ability to photosynthesize. All photosynthetic producers harvest the energy of the sun and create food for the consumers in an ecosystem. In addition, the photosynthetic producers consume the carbon dioxide animals release as waste products. In the process of photosynthesis light gathering pigments absorb photons from the sun. These photons cause electrons in the pigments to become excited. The electrons will ultimately give up their energy in an electron transport system resulting in the production of ATP. Chlorophyll a is the most common photosynthetic pigment, but a wide range of additional pigments have been described. Different types of pigments preferentially absorb different wavelengths of light, and as such, different pigment containing organisms are found in diverse environments.

In this activity, you will investigate photosynthetic bacteria isolated from a Winogradsky column. The Winogradsky column is named for a famous Russian microbiologist, Sergei Winogradsky. These columns are created by packing a long glass cylinder with pond mud, and then adding carbon substrates (hay, newspaper, ground meat) and pond water. The cylinder is covered with aluminum foil and then exposed to light for several weeks. Oxogenic and anoxygenic populations of photosynthetic bacteria will establish themselves at various locations throughout the column. Different populations of bacteria can be easily differentiated on the basis of color due to photosynthetic pigment production.

CHECK FOR URL UPDATES

Before beginning this activity, check on the Wadsworth Biology Resource Center homepage for URL updates and information:

http://www.wadsworth.com/biology

Once you have arrived at the Wadsworth Biology Resource Center, select General Biology and locate this book. Any updates should appear.

ACTIVITY

1. Using Netscape or another Web browser, type the following address at the URL prompt:

    http://www.mannlib.cornell.edu:10000/ccr2/index.html

2. Select the Energy button.

3. Select Photosynthesis-an overview.

4. Read all of the information contained on the Energy pages

5. Formulate an opinion as to why the different types of bacteria segregate themselves vertically in the Winogradsky column. Record this information on the activity report sheet.

6. Select Why not test your hypothesis experimentally?

7. Read the scenario presented, paying particular attention to the absorption vs. wavelength curve presented for different chlorophyll types. (you may prefer to print this graph for future reference)

8. Select Design your experiment, then click here.........

9. Select the name of each bacterium one by one to see the wavelength at which the bacterium's pigment preferentially adsorbs light. When you select the name of the bacterium, the hand will release the organism which will then travel to a particular location on the spectrum. Watch closely and quickly record your observations. If the bacterium disappears before you have recorded the wavelength, simply select the bacterium again to repeat the process.

10. When you have decided which bacterium contains chlorophyll a, select HERE.

11. Enter your results to see if you have chosen correctly.

12. Answer the questions on the Activity Worksheet.

PHOTOSYNTHESIS ACTIVITY WORKSHEET

Name _____ Date _____

Class/Section _____

1. What is the term used to describe bacteria which can convert light energy to ATP?

2. Complete the table below:

Type of bacterium	Wavelength at which pigment best absorbs light
1.	
2.	
3.	
4.	

3. Which bacterium contains chlorophyll a?

8

STEP-BY-STEP GLYCOLYSIS

INTRODUCTION
Organisms stay alive by taking in food. This food is then broken down in catabolic pathways to provide cells with energy to carry out cellular activities. The energy currency in living systems is adenosine triphosphate (ATP). Therefore, the objective of catabolic processes is to produce ATP. The ATP can then be utilized by the cell for growth and repair. The breakdown of glucose—glycolysis—is the first step of the major energy releasing pathways. This reaction takes place in the cytoplasm of both procaryotic and eucaryotic cells.

Glucose is a six-carbon sugar. In the process of glycolysis, glucose is converted to two, three-carbon molecules of pyruvate. This breakdown is accomplished through a series of events including phosphorylation, dehydration, oxidation, and reduction. In the absence of oxygen, pyruvate can be further converted to lactate. The conversion of pyruvate to lactate often takes place in the muscles during vigorous exercise as oxygen in the tissues becomes depleted.

CHECK FOR URL UPDATES
Before beginning the activities in this chapter, check Wadsworth Biology Resource Center home page for URL updates and information:

http://www.wadsworth.com/biology

Once you have arrived at the Wadsworth Biology Resource Center home page, select General Biology and locate this book. Any updates should appear.

ACTIVITY

While it is possible to simply memorize the reactions in glycolysis, in order to understand the process, it is important to know why each step takes place as it does. The site you will use for this activity is the DIY Glycolysis site maintained by the Department of Biochemistry at the University of Leeds in the United Kingdom. In the activity you will choose the individual reactions in the conversion of glucose to lactate. If you choose an incorrect, illogical, or impossible reaction, the site will provide you with hints as to the correct selection. If you choose the correct reaction, you will be directed to the next step.

It is very important that you have a basic understanding of the steps in glycolysis before you begin this activity. In addition, it is important to understand the processes of phosphorylation, oxidation, reduction, and the generation of ATP. The DIY site also has a link available for reviewing glycolysis.

PART 1

1. Using Netscape, or another Web browser, type in the following address at the URL prompt:

    http://bmbwww.leeds.ac.uk/designs/diygly/home.htm

2. Select Step by Step Glycolysis if you need to review the process.

PART 2

1. Select Begin with Glucose.

2. The page will prompt you to answer the question, "Why are you using this site?"

3. Select I'm required to work on it as part of a course or another applicable reason.

4. Select Glucose.

5. In each subsequent step select the correct reaction type from the list. If you choose incorrectly, read the complete explanation on the screen before using the "Back" key of your Web browser to return to the previous page. The explanations of incorrect selections will help you choose the correct reaction. If you choose the correct reaction, scroll back up to the top of the page and choose the Next step. Throughout each step the number of ATPs produced or spent will be tallied at the bottom of each page.

6. Continue working through the selections until you have successfully converted glucose to lactate. *Note:* It is helpful to have available a diagram of the molecule lactate in front of you as you work through this series of reactions. Knowing the ultimate structure of the molecule can make your reaction selections easier.

P A R T   T W O

PRINCIPLES OF INHERITANCE

9

MITOSIS

INTRODUCTION
Cells of the human body must replicate in order to replace dead cells, repair damaged or injured tissues or allow the body to grow. Cells of the gastrointestinal tract and the skin replicate throughout one's life, liver cells only replicate intermittently, while nerve cells do not replicate at all. The process of replication is tightly controlled and very precise. It is imperative that each dividing cell produce two daughter cells genetically identical to the parent cell. Somatic cell division, or mitosis, is the process by which cells divide. For convenience, mitosis is divided into four parts, prophase, metaphase, anaphase, and telophase. The entire process takes on average, one hour to complete.

In this activity, you will participate in an on-line interactive review of the process of mitosis. Before beginning the activity, you should have a complete understanding of significant events which take place during the different phases of mitosis. Review the material in your text book if you have not mastered the subject. Once at the site, you will be provided with photomicrographs of mitotic plant or animal cells. You will be asked to identify each stage of mitosis and answer questions particular to that phase. The site will provide you with feedback on your answers.

CHECK FOR URL UPDATES
Before beginning this activity, check on the Wadsworth Biology Resource Center homepage for URL updates and information:

http://www.wadsworth.com/biology

Once you have arrived at the Wadsworth Biology Resource Center homepage, select General Biology and locate this book. Any updates should appear.

ACTIVITY 1

1. Using Netscape or another Web browser, type the following address at the URL prompt:

   http://biog-101-104.bio.cornell.edu/BioG101_104/tutorials/cell_division.html

2. Select <u>Whitefish mitosis</u> to review animal mitosis

3. Select <u>Onion Root Tip</u> to review plant mitosis.

4. Select <u>Question 1</u>. Answer the question(s) by using the selections in the pull down menus.

5. After making your selection, <u>select OK</u>. The correctness of your answer will immediately be evaluated. If you answer the question incorrectly, continue to choose answers until you have answered correctly.

6. Continue to select and answer all questions (1-6) as described in steps 4 and 5.

7. Use the Mitosis Activity Report Sheet to provide an explanation for all questions you initially answered incorrectly.

ACTIVITY 2

1. If you are interested in visiting a virtual mitosis site, type in the following address at the URL prompt:

   http://www.biology.uc.edu/vgenetic/mitosis/mitosis.htm

2. The graphics will take a short time to load, but once the site is fully loaded, the photomicrographs will be shown sequentially. This site also provides a glossary of mitotic terms.

MITOSIS ACTIVITY WORKSHEET

Name _____   Date _____

Class/Section _____

1. In the space below, list any phase of mitosis you initially identified incorrectly. Describe in detail the events which occur during this particular phase. Include an explanation as to why you made the initial choice incorrectly.

10

MEIOSIS REVIEW

INTRODUCTION
When producing gametes, through the process of meiosis, it is imperative that the genetic material be replicated and appropriately partitioned into daughter cells. Unlike mitosis, meiosis consists of two cell divisions ultimately resulting in four haploid daughter cells. In males, all four of these cells will be functional sperm whereas in females, three of the daughter cells will be nonfunctional polar bodies, and one of the cells will be a functional egg. The reduction of genetic material from diploid to haploid takes place in meiosis I when maternal and paternal members of homologous pairs separate. The second division of meiosis occurs very much like mitosis. When egg and sperm are united in a fertilization event, the diploid number of chromosomes will be recreated in the zygote.

In this activity, you will take part in an interactive review of meiosis. This review will involve answering questions as well as interpretation of graphics. Not only will the review test your knowledge of meiosis but it will also expect you to make comparisons between meiosis and mitosis. It is suggested that you review the material and schematics related to meiosis in your text before beginning this exercise.

CHECK FOR URL UPDATES
Before beginning this activity, check on the Wadsworth Biology Resource Center homepage for URL updates and information:

http://www.wadsworth.com/biology

Once you have arrived at the Wadsworth Biology Resource Center homepage, select General Biology and locate this book. Any updates should appear.

ACTIVITY

1. Using Netscape or another Web browser, type the following address at the URL prompt:

    http://biog-101-104.bio.cornell.edu/BioG101_104/tutorials/cell_division.html

2. Under the heading "Name, identify, and describe important events in meiosis", select Review.

3. Review the graphics paying special attention to the appearance, location and number of chromosomes.

4. Use the "Back" key of your browser to return to the previous page.

5. Select Question 1

6. Answer the question using either the pull down menu or by typing the appropriate answer.

7. You will be provided with immediate feedback on your answer. Provide an explanation for any wrong answer on the activity worksheet.

8. Continue to work through all five questions by repeating steps 5, 6, and 7.

9. Use the "Back" key of your browser to return to the BioG 101-104 homepage.

10. Under the heading "Describe important similarities and differences between mitosis and meiosis", select Question 1.

11. Use your mouse to make the appropriate check box selections (Note you may have to leave all check boxes empty to correctly answer a question).

12. Continue to work through Question 2 using steps 9, 10, and 11 of this activity. Provide an explanation for any wrong answer on the meiosis activity worksheet.

13. Use the "Back" key of your browser to return to the BIOG 101-104 homepage.

14. Under the heading "Differentiate between the following pairs of terms..." select Question 1. Answer all six questions as previously described in this activity.

15. Provide an explanation for any incorrect answers you provided on the activity worksheet.

MEIOSIS REVIEW ACTIVITY WORKSHEET

Name _____ Date _____

Class/Section _____

1. Provide an explanation for incorrect answers to questions under "Name, identify and describe   important events in meiosis."

2. Provide an explanation for incorrect answers to questions under "Describe important similarities and differences between mitosis and meiosis".

3. Provide an explanation for incorrect answers to questions under "Differentiate between the following pairs of terms..."

11

MENDELIAN GENETICS AND THE VIRTUAL FLY LAB

INTRODUCTION

Gregor Mendel established the foundation of genetic principles for the field of transmission genetics. These principles, derived directly from experiments Mendel carried out with the garden pea, describe how genes are transmitted from parent to offspring. At the time Mendel carried out his experiments, there was no knowledge of the existence of chromosomes or the nature of the genetic material. Nonetheless, Mendel was able to determine that distinct units of inheritance (genes) are transmitted in a regular and predictable pattern. Mendel's initial experiments included crosses involving only one pair of contrasting traits. From these experiments he was able to determine that inherited units existed in pairs and that one member of the pair was dominant over the other. When Mendel performed monohybrid crosses with true breeding plants, his results were always the same. In the F_1 generation, all plants displayed the dominant phenotype. In the F_2 generation, three out of four plants displayed the dominant phenotype with one plant showing the recessive trait. Mendel was fortunate in that he selected traits that happened to be on seven different chromosomes. His principles are the cornerstone of all work done in genetic research labs today.

A common experimental animal used in many genetics research laboratories is the fruit fly *Drosophila melanogaster*. Many different mutants are available, with flies being easily grown and manipulated in the laboratory.

CHECK FOR URL UPDATES

Before beginning the activities in this chapter, check the Wadsworth Biology Resource Center home page for URL updates and information:

http://www.wadsworth.com/biology

Once you have arrived at the Wadsworth Biology Resource Center home page, select General Biology and locate this book. Any updates should appear.

ACTIVITY

In this activity you will access the Virtual Fly Lab, a project of California State University at Los Angeles. You can design experimental crosses and generate and analyze genetic data of your own through selection of gene types and design of various matings. Virtual Fly Lab can be used to demonstrate Mendelian ratios, recessive and dominant phenotypes, autosomal and sex linkage, gene lethality, and map distances.

The objective of this activity is to design several crosses to illustrate various genetic principles. Before you begin the activity, there are a few guidelines to follow and characteristics of the site that you should be familiar with.

- You can select genetic characteristics by using the radio buttons in the listing of genetic traits.
- Wild type is the default condition. One member of the mating pair must remain a wild type fly. It is only possible to select a mutant characteristic for either the male or the female, not both.
- When you select a mutation, the fly will be homozygous for that mutation, unless the mutation is lethal. Lethal mutations are heterozygous.
- You have the option of selecting mutations for nine different characteristics.
- A glossary listing all gene designations is included at the end of the test cross results.

The Virtual Fly Lab is an activity that should only be undertaken once you have a good grasp of Mendelian genetics.

PART 1

1. Using Netscape, or another Web browser, type the following address at the URL prompt:

   http://vflylab.calstatela.edu/edesktop/VirtApps/VflyLab/IntroVflyLab.html

2. Select "Design a Cross Between Two Flies"

3. Scroll down through the nine genetic characteristics. Select a mutation for only one characteristic.

4. Select the "Mate flies" button.

5. Use the report sheet at the end of the activity to record the results of your cross.

6. Scroll down the page until you come to the heading "Other Places to Go."

PART 2

1. Under the heading "Other Places to Go," select <u>Design and Mate Flies</u>.

2. Select two mutations in two of the nine genes.

3. Select "Mate Designed Flies."

4. Record the results of your dihybrid cross on the Mendelian Genetics and the Virtual Fly Lab Worksheet at the end of the chapter.

PART 3

1. Select <u>Design flies</u> at the bottom of the page.

2. Select mutations in three of the nine genes.

3. Select "Mate Designed Flies."

4. Record the results of the trihybrid cross on the Mendelian Genetics and the Virtual Fly Lab Worksheet at the end of the chapter.

PART 4

If you'd like to read Mendel's original paper (in English or German) in which he describes his experiments, as well as comments on his paper, type in the following address at the URL prompt:

http://hermes.astro.washington.edu:80/mirrors/MendelWeb/

Follow the directions from your instructor regarding the submission of your worksheets. Some instructors will want them e-mailed; others will want them handed in.

MENDELIAN GENETICS AND THE VIRTUAL FLY LAB WORKSHEET

Name _____ Date _____

Class/Section _____

PART 1: MONOHYBRID CROSS

1. What are the phenotypes of the parents?

2. What are the phenotypes of the offspring?

3. Was the mutation you selected dominant or recessive?

4. What was the expected phenotypic ratio of the offspring?

5. What was the actual phenotypic ratio of the offspring?

PART 2: DIHYBRID CROSS

1. What are the phenotypes of the parents?

2. What are the phenotypes of the offspring?

3. Were the mutations you selected dominant or recessive?

4. What was the expected phenotypic ratio of the offspring?

5. What was the actual phenotypic ratio of the offspring?

PART 3: TRIHYBRID CROSS

1. What are the phenotypes of the parents?

2. What are the phenotypes of the offspring?

3. Were the mutations you selected dominant or recessive?

4. What was the expected phenotypic ratio of the offspring?

5. What was the actual phenotypic ratio of the offspring?

6. Were any of the mutations you selected lethal?

12

EXPLORING THE HUMAN GENOME MAP

INTRODUCTION

The Human Genome Project is an international research effort with several goals including mapping and sequencing of the entire human genome. A byproduct of genome investigation has been the identification and mapping of over 1000 human disease genes. Each October the most recent maps of the 23 human chromosomes, including locations of disease genes, are published in the journal *Science*. In addition, these maps can be viewed using the world wide web by accessing *Science On-line*.

The large amounts of data being generated by the Human Genome project are organized and made available to scientists and the general public via several data bases. On-line Mendellian Inheritance in Man (OMIM) is an example of one database. OMIM is a catalog of human genes and genetic disorders developed for the World Wide Web by the National Center for Biotechnology Information. The database contains information on the research history of the disease gene including extensive references, a clinical synopsis of the disorder, inheritance patterns, and information on the molecular nature of the mutation or defect leading to the disorder.

In this activity, your objectives will be to access the most recent map of the human genome, select one recently mapped disease gene, and research the details of the disease gene using Online Mendellian Inheritance in Man.

CHECK FOR URL UPDATES

Before beginning this activity, check on the Wadsworth Biology Resource Center homepage for URL updates and information:

http://www.wadsworth.com/biology

Once you have arrived at the Wadsworth Biology Resource Center homepage, select General Biology and locate this book. Any updates should appear.

ACTIVITY 1

1. Using Netscape or another Web browser, type the following address at the URL prompt:
    http://www.ncbi.nlm.nih.gov

2. Select "Gene Map of the Human Genome".

3. Read the introduction which describes the type of mapping and technologies used to generate the map of human disease genes.

4. Answer the questions regarding mapping in the Activity Report Sheet.

ACTIVITY 2

1. Select one of the 23 chromosomes (Browse by Chromosome).

2. Read the short paragraph of information regarding one of the disease genes.

3. Select a disease gene.

4. Read the information regarding your disease gene and answer the questions in the activity report sheet.

ACTIVITY 3

1. Select the OMIM entry for your gene (Database Records).

2. Answer the questions regarding this gene in the activity report sheet.

EXPLORING THE HUMAN GENOME MAP ACTIVITY WORKSHEET

Name _____   Date _____

Class/Section _____

ACTIVITY 1

1. What type of chromosome map are you viewing?

2. Define the following terms:

    cDNA

    STSs

3. What percentage of the human genome actually consists of coding sequences?

Exploring the Human Genome Map   47

ACTIVITY 2

1. Record the following information:

   Chromosome selected:

   Gene selected:

   Normal function of the gene product:

   Clinical manifestations:

   Incidence:

   Treatment:

EXPLORING THE HUMAN GENOME MAP, CONT.

ACTIVITY 3

1. What is the gene map locus of this disease?

2. Are there any alternative names for this genetic disorder?

3. What is the molecular nature of the genetic mutation?

4. What is the mode of inheritance of this particular disorder?

13

ETHICAL, LEGAL, AND SOCIAL IMPLICATIONS OF THE HUMAN GENOME PROJECT

INTRODUCTION

As sequence information from the Human Genome Project continues to accumulate, the genes associated with various genetic diseases are being identified, making it possible for physicians to presymptomatically diagnose disease and offer either preemptive treatments or a greater range of reproductive choices. With testing currently available for Cystic Fibrosis, Fragile X Syndrome, Gaucher's disease, Retinoblastoma, and others, at-risk individuals can be screened and appropriate decisions can be made. In addition to affecting the areas of disease diagnosis and treatment, the Human Genome Project will revolutionize the approach to health care coverage. Use of genetic data by insurance companies to deny coverage or charge excessive rates will effectively void any good that can potentially be derived from genetic research. It is the charge of the National Institutes of Health–Department of Energy Working Group on Ethical, Legal, and Social Implications (ELSI) of the Human Genome Project to formulate a series of recommendations for local and federal policy makers such that broad discrimination in health care coverage due to family genetic history will not become a reality.

CHECK FOR URL UPDATES

Before beginning the activities in this chapter, check the Wadsworth Biology Resource Center home page for URL updates and information:

http://www.wadsworth.com/biology

Once you have arrived at the Wadsworth Biology Resource Center home page, select General Biology and locate this book. Any updates should appear.

ACTIVITY

In this activity you will access the ELSI Scope Note home page. On this page you will find Scope Notes on various topics including human gene therapy and genetic testing. Scope Notes are produced at the National Reference Center for Bioethics Literature, Kennedy Institute of Ethics, Georgetown University, Washington, DC. These notes provide an overview of issues concerning controversial subject matter related to specific topics in biomedical ethics. Of special interest are the notes related to the Human Genome project.

In this activity, your first objective is to read Scope Notes 22 and 24. Formulate an opinion regarding testing, therapy, and insurance. Your second objective is to take part in an online discussion of the use of genetic data by health insurance companies on the home page maintained by the journal Science.

PART 1

1. Using Netscape, or another Web browser, type in the following address at the URL prompt:

   http://www.ncgr.org/elsi/scopenotes.hp.html

2. Select Educational Resources, and then select Scope Notes Series.
3. Formulate an opinion regarding gene therapy and genetic testing.
4. Answer the questions on the Ethical, Legal, and Social Implications of HGP Worksheet at the end of this chapter.

PART 2

1. Using Netscape, or another Web browser, type in the following address at the URL prompt:

   http://www.aaas.org/science/science.html

2. Select Full text.
3. Click on the "Search All Issues" icon.
4. Once on the search page enter the following data:

Kathy Hudson	Author
Genetic Discrimination	Words in Title
391	First Page

5. Once the title of the author's article appears, select Summary. After selecting Summary you will now be required to register touse Science ONLINE. It is possible to register for free partial access. If you wish to register, follow the instructions on the screen and provide the appropriate information. Once you have registered, you will have to go back to Step 3, and then continue with Steps 4 and 5. When you request the article summary, you will now be asked to sign in. Enter your user name and password and then you will be given access to the requested article.
6. Once the article has appeared, read the article.
7. Select Go to the Discussion, and click on the "Post a New Message" button.

52 Internet Activities for General Biology

8. Add the appropriate information (see below) and then your opinion.

   Enter Your Name Below:

   Enter Your E-mail Address Below:

   Enter Message Subject Below:

   Enter Message Below:

9. Select <u>Submit This Message</u>.

PART 3
1. Monitor the discussion in the Science forum for one week. Print any responses to your opinion.
2. On the due date of the assignment, turn in your typewritten answers to the Scope Note questions and any printed responses to the interactive dialogue on the Science home page.

Follow the directions from your instructor regarding the submission of your worksheets. Some instructors will want them e-mailed; others will want them handed in.

ETHICAL, LEGAL, AND SOCIAL IMPLICATIONS OF HGP WORKSHEET

Name _____ Date _____

Class/Section _____

1. Should insurance companies be informed of genetic disease that runs in a family?

2. Should these companies be permitted to charge higher rates to individuals who have a family history of genetic disease?

3. Should any genetic defects of the fetus be considered a preexisting condition?

4. If one family member manifests a genetic illness, such as Huntington's disease, should the rest of the siblings be forced to be tested or lose their health care coverage?

5. Should insurance companies pay for expensive gene therapy procedures?

6. Can you think of any state or federal legislation that should be enacted so that genetic discrimination cannot take place?

14

HOW MUCH DO WE NEED TO KNOW? ETHICAL CASES IN GENETICS

INTRODUCTION
In the spring of 1902, Walter Sutton, came to a revolutionary idea that represented a major breakthrough in the study of biology. The idea was simple: that the Mendelian processes of segregation and independent assortment were explainable if genes were located on chromosomes. This explained many events associated with the transmission of traits through generations. His idea is central to the study of modern biology. Its importance can be seen in the formation of the Human Genome Project in October 1990, which has as its mandate both the identification and locating of the approximately 100,000 genes carried on the human chromosome set.

Although the technological aspects of the Human Genome project seem to receive most of the media's attention, it is the moral, legal, and ethical fallout from this massive accumulation of genetic information that should be considered far more seriously than it is currently. We are usually quick to condemn big business in its misuse of this type of data, but how would you use this type of information if you were a genetic counselor? What would you do if one of your clients tested positively for a negative, dominant genetic trait? How would you feel about telling your client this horrible news? Would you be obligated, either morally or legally, to tell his or her children, or, perhaps, even his or her spouse? How would you personally respond to the announcement that you, or one of your parents, had tested positive to a particular genetic defect that had a probability of being inherited approximately 50 percent of the time?

CHECK FOR URL UPDATES
Before beginning the activities in this chapter, check the Wadsworth Biology Resource Center home page for URL updates and information:

http://www.wadsworth.com/biology

Once you arrive at the Wadsworth Biology Resource Center home page, select General Biology and locate this book. Any updates should appear.

ACTIVITY

In this activity, you will visit the University of Kansas Medical Center home page for genetics professionals. This site links to the University's Genetics Education Center page. From the Education Center page, you will access two different genetic disease scenarios, read the cases, and then comment on the situation. The cases request that you post a comment; however, we request that you do not post one as a courtesy to genetics professionals who regularly access this site. Instead, please take part in the polling process listed at the end of the posting section.

PART 1

1. Using Netscape, or another Web browser, type the following address at the URL prompt:

    http://www.kumc.edu/gec/geneinfo.html

    You are now on the Kansas University Medical Center Information for Genetic Professional's home page.

2. Scroll down the page until you come to Genetics Education Center. Select this phrase.

3. From the Genetics Education Center page, search under the Human Genome Project section for "Ethics cases from the Exploratorium exhibit on genetics."

4. Select Huntington's disease scenario. Read the case. Record your answers to the questions and compare your answers with the rest of your class.

5. Select Click here to see the results of other people's responses.

PART 2

After reading the case, answer the questions on the Huntington's Disease Case Scenario Worksheet at the end of this chapter.

PART 3

Return to the Genetics Education Center home page and repeat the same steps you took for the previous activity but select Cystic fibrosis scenario. After reading the case, answer the questions on the Cystic Fibrosis Case Scenario Worksheet at the end of this chapter.

Follow the directions from your instructor regarding the submission of your worksheets. Some instructors will want them e-mailed; others will want them handed in.

58 Internet Activities for General Biology

HUNTINGTON'S DISEASE CASE SCENARIO WORKSHEET

Name _____ Date _____

Class/Section _____

1. Does George's right to know and be tested for the Huntington's disease gene supersede his father's wish not to know his own genetic status?

2. Does George's father, Mr. F., have a right not to know?

3. Would you want to know your genetic status for an autosomal dominant disease gene?

4. How do your responses to the poll compare with the responses of other people? (Were you in the majority or minority?)

CYSTIC FIBROSIS CASE SCENARIO WORKSHEET

Name _____   Date _____

Class/Section _____

1. Should the genetics counselor tell *both* parents about the test results?

2. Should Mr. C. be told that he is definitely not the father of the child that was tested?

3. The baby's biological father must be a carrier of the recessive trait. Should the genetics counselor attempt to identify and locate the father and inform him of the potential risk to his future offspring?

4. How do your responses to the poll compare with the responses of other people? (Were you in the majority or minority?)

15

MOLECULAR BIOLOGY VOCABULARY

INTRODUCTION
Within the broad discipline of biology there are many subdisciplines that use the same vocabulary to describe concepts and processes. However, in those same subdisciplines, researchers and lecturers frequently use jargon or slang to describe a particular activity, action, or substance. In no area is this more common than in the area of molecular biology. In particular, molecular biologists will frequently use acronyms to describe a particular technique. These acronyms, as well as other terms, have come into common usage and are now accepted as a normal part of the vocabulary of the discipline.

CHECK FOR URL UPDATES
Before beginning the activities in this chapter, check the Wadsworth Biology Resource Center home page for URL updates and information:

http://www.wadsworth.com/biology

Once you have arrived at the Wadsworth Biology Resource Center home page, select General Biology and locate this book. Any updates should appear.

ACTIVITY
The Human Genome Program has put together a *Primer on Molecular Genetics* that not only describes various concepts and techniques, but also includes a glossary of molecular biology terms. In this activity you will access the glossary to find the definitions of several terms and then make use of the primer to find out in what context these terms are most frequently used.

PART 1

1. Using Netscape, or another Web browser, type in the following address at the URL prompt:

   http://www.gdb.org/Dan/DOE/intro.html

2. Once on this page Select Glossary.

3. Use the glossary to find the definitions for the following eight terms:
   PCR   RFLP
   STS   ETS
   cDNA   cosmid
   rare enzyme cutter tandem repeat sequences

4. Record the definitions of these terms on the Molecular Biology Vocabulary Worksheet at the end of this chapter.

PART 2

1. Return to the contents page of the primer and then use other headings to find information on techniques or applications associated with four of the previous terms.
2. Record this information on the Molecular Biology Vocabulary Worksheet.

Note: It may be difficult to sort through the primer without a hard copy of the document. In that case you can access the primer through another URL that is maintained by the Human Genome Management Information System. Once at the Human Genome project site, you can then either download the primer onto a disc or print the primer (it's forty-one pages, so please do this on either your own printer or on a Network printer only during a slow period of the day).

PART 3, OPTIONAL

1. To access the primer through the HGP home page, type in the following address at the URL prompt:

   http://www.ornl.gov/TechResources/Human_Genome/home.html

   Once on the home page, look under "General Information."

2. Select the Primer on Molecular Genetics.

3. Follow the directions regarding use of the Acrobat reader to access or download the document.

Follow the directions from your instructor regarding the submission of your worksheets. Some instructors will want them e-mailed; others will want them handed in.

MOLECULAR BIOLOGY VOCABULARY WORKSHEET

Name _____ Date _____

Class/Section _____

1. Provide definitions for each of the following terms:

   PCR

   RFLP

   STS

   ETS

   cDNA

   cosmid

   rare enzyme cutters

   tandem repeat sequences

2. Describe a technique or application associated with four of the molecular biology vocabulary words you previously defined.

16

DNA ISOLATION

INTRODUCTION

At its most basic level the double-stranded DNA molecule can be visually represented as a ladder with the sugar-phosphate molecules forming the backbone or sides of the ladder and the nitrogenous bases forming the interior or rungs of the ladder. In its linear form, the molecule of a tiny procaryotic cell measures approximately 1mm in length. Because the typical bacterial cell is only 1 to 4μm in length, the DNA must assume a secondary structure in order to fit inside the bacterial cell. The DNA ladder does indeed twist, forming the classical double helix. The molecule then supercoils back on itself so that ultimately the DNA is condensed tightly enough that it will fit into a bacterial cell or the nuclei of a eucaryote.

The size and solubility properties of the DNA molecule make it easy to isolate relatively intact. Isolation of DNA involves lysing the cell wall (if the organism has one), disrupting the cell membrane, precipitating and removing cellular proteins, and then finally precipitating the DNA. Although one might expect DNA isolation to involve the use of expensive scientific equipment and chemicals, it actually can be done using a simple detergent, test tubes, ethanol, and a glass rod.

CHECK FOR URL UPDATES

Before beginning the activities in this chapter, check the Wadsworth Biology Resource Center home page for URL updates and information:

http://www.wadsworth.com/biology

Once you have arrived at the Wadsworth Biology Resource Center home page, select General Biology and locate this book. Any updates should appear.

ACTIVITY

In this activity you will access a set of experimental protocols. Depending on the organization of your course and the time available, you may be able to include several of these experiments into your class time.

In this activity you will use an experimental protocol found on the Access Excellence site that will allow you to isolate DNA from onion. This protocol can be carried out individually, in groups, or as a demonstration by your instructor. It is not necessary to have an actual laboratory period in which to perform this activity. If it is simply not possible to carry out the experiment at all, it is still important to read the protocol. You will see that the procedure requires no sophisticated equipment or technology, but is simply a use of common chemicals that take advantage of the physical properties of the large viscous DNA molecule. If you and/or your class carry out the procedure, it will take approximately one hour to complete. To speed up the isolation process, your instructor may want to prepare the homogenate before the class period.

In addition to the DNA isolation experiment, several other procedures can be found at this site. You can print these experiments and have them available for use later in the term.

1. Using Netscape, or another Web browser, type in the following address at the URL prompt:

    http://www.gene.com/ae/

2. Select <u>Activities Exchange</u>.
3. Select the <u>Woodrow Wilson Biology Collection</u> (found under the "AE Partners Collection" heading).
4. Select <u>1994 Populations to Molecules</u>.
5. Select <u>Extraction of DNA from Onion</u> (under the heading "IV. Molecular Genetics").
6. Print or download the protocol onto a disc.
7. Before carrying out the procedure, read the steps of the experiment carefully and then answer the questions found on the DNA Isolation Activity Worksheet found at the end of this chapter.

Note: Access Excellence is a national education program sponsored by Genetech Inc. The project aims to bring industry, scientists, and public educators together for an exchange of scientific information that can then be introduced into the classroom. The Access Excellence site has many useful resources and hyperlinks. Science in the news, biotech information, activities exchange, and a listing of biological sites of interest on the Web are but a few of the topics to be found. Be sure to explore this site further when you have the time.

Follow the directions from your instructor regarding the submission of your worksheets. Some instructors will want them e-mailed; others will want them handed in.

DNA ISOLATION ACTIVITY WORKSHEET

Name _____ Date _____

Class/Section _____

1. Why must you wear gloves when performing this protocol?

2. Following homogenization of the onion in a blender and filtration of the homogenate through cheesecloth, what macromolecules are present in your homogenate filtrate?

3. Why is ice cold ethanol added to the homogenate?

4. What physical properties of DNA allow it to be spooled onto the glass rod?

17

TRANSCRIPTION AND TRANSLATION

INTRODUCTION

At its most basic level, DNA is composed of nitrogenous bases, phosphate, and ribose molecules. The sequential ordering of the bases will ultimately encode the form and function of a living organism. But, how can a simple sequence of bases can give rise to such a complex organism? In order for the DNA sequence to be expressed it must first be converted into a sequence of RNA and ultimately the RNA sequence will be converted into a chain of amino acids. This linear sequence of amino acids is the primary structure of a protein. Proteins have many different functions in the cell including, catalysis, structural activities, receptors, storage, protection, motion, and transport. Converting the linear base sequence of DNA into RNA is known as transcription, and converting the RNA into protein is a process called translation.

For this activity, you'll be completing a quiz in molecular biology on the Molecular Biology Web Exercise homepage. The quiz consists of 46 questions on various topics including DNA and RNA structure, transcription, translation, restriction enzymes, the genetic code and RNA processing. Some of the questions will require you to choose an answer, while other questions can be answered by typing words in the blank box. Before beginning the quiz, it is important that you have an understanding of the topics previously listed. Review the material in your text book to prepare for the quiz. If you need additional information or would like to see an animation on the topic of translation, the material in Activity 2 will be helpful.

CHECK FOR URL UPDATES

Before beginning this activity, check on the Wadsworth Biology Resource Center homepage for URL updates and information:

http://www.wadsworth.com/biology

Once you have arrived at the Wadsworth Biology Resource Center homepage, select General Biology and locate this book. Any updates should appear.

ACTIVITY 1

1. Using Netscape or another Web browser, type the following address at the URL prompt:

   http://www.medkem.gu.se:80/ln/molbio/gene/

2. Select This One requires Netscape 2.0/3.0

3. Begin answering the questions. If you are presented with a blank text box, place your cursor in the box, click, and then begin typing. If you are presented with a check box, click the appropriate box to answer the question. After you have typed or selected your answer, select the Check button. You will be provided with immediate feedback concerning your answers. You will not be permitted to move to the next question until the previous question is answered correctly.

4. Continue answering all 46 questions.

ACTIVITY 2

1. If you'd like to review additional material on the subject of translation, type the following address at the URL prompt:

   http://dna.chem.rochester.edu/protein_synth/translation.html

2. Select Proteins: An Introduction. Review the material.

3. Use the "back" key of your browser to return to the previous page.

4. Select Protein Synthesis Overview. Review the material

5. Use the "back" key of your browser to return to the previous page.

6. Select Protein Synthesis: Animation. The animation will take a short time to load, but once its loaded, you will be able to play the animation repeatedly by clicking on the ribosome.

Note on Activity 1:

Some parts of this quiz are suitable only for the advanced student. Some of the questions are difficult and require a strong background in biochemistry. Several of the chemical structures are presented as three dimensional space filling molecules. These may be viewed with either Rasmol or Kinemage (downloadable at the site) or as flat images without the software. Several questions require the use of a sequence data base. If you have not used a data base before, the following tips may prove helpful: 1) to search for your particular sequence, simply copy the sequence provided to the clipboard and then paste the sequence in the query box once you are at the sequence search site, and 2) be sure you are searching the correct database (amino acid sequence vs. nucleotide sequence). This quiz can take up to one hour to complete.

18

CANCER: A LOSS OF CELL CYCLE CONTROL

INTRODUCTION

Cellular replication, or the cell cycle, is controlled and highly regulated by internal and external chemical signals. Occasionally cells arise that no longer recognize or respond to these regulatory signals. The cells that ignore chemical messages and divide uncontrollably are referred to as being transformed. Transformed cells specialize in replication but do not differentiate into the appropriate cell type for the organ in which they are proliferating. When growing, transformed cells pile on top of each other and ignore signals from adjacent cells. This loss of contact inhibition is an important characteristic of a transformed cell. Most transformed cells are located and destroyed by the immune system. If a transformed cell is not destroyed, it will continue to replicate uncontrollably and eventually form a solid mass, or tumor. Tumor cells actively encourage blood vessels in the local area to grow and vascularize, thereby providing the tumor with its own supply of food and oxygen. Tumors that rapidly grow and invade surrounding tissue are described as malignant, whereas tumors that are self-contained and noninvasive are described as benign. Malignant tumors overwhelm normal cells as they overgrow and displace normal tissues. If untreated, these tumors will ultimately cause loss of organ function and eventual death of the host due to a depletion of tissue nutrients as well as the inability of the transformed cell to carry out normal tissue or organ function. The group of diseases characterized by malignant tumors is referred to collectively as cancer.

CHECK FOR URL UPDATES

Before beginning the activities in this chapter, check the Wadsworth Biology Resource Center home page for URL updates and information:

http://www.wadsworth.com/biology

Once you have arrived at the Wadsworth Biology Resource Center home page, select General Biology and locate this book. Any updates should appear.

ACTIVITY

In this activity you will access the University of Pennsylvania's multimedia information page on oncology. This page contains information on cancer types, diagnoses, and therapies. It also provides access to press releases related to cancer advancements as well as personal stories of cancer survivors. Many of the articles found at this site are written for the patient, or lay person, and others are written for physicians. In Part 1 of this activity, you will learn the basic facts about cancer. In Part 2 you will read a patient's personal story about his battle with mesothelioma, a type of lung cancer frequently caused by asbestos exposure.

PART 1

1. Using Netscape, or another Web browser, type in the following address at the URL prompt:

   http://cancer.med.upenn.edu/

2. Select Disease-Oriented Menus.

3. Select General Information about Cancer.

4. Select The Nature of Cancer (patient oriented).

5. Read the article and answer the questions on the Cancer Worksheet at the end of this chapter.

PART 2

1. Use the "Back" key of your Web browser to return to the Disease-Oriented Menus.

2. Scroll down the page to Adult Cancers.

3. Select Lung Cancer.

4. Select Mesothelioma.

5. Select Thoughts from a Mesothelioma Patient.

6. Answer the questions found on the Cancer Worksheet at the end of this chapter.

Follow the directions from your instructor regarding the submission of your worksheets. Some instructors will want them e-mailed; others will want them handed in.

CANCER WORKSHEET

Name _____ Date _____

Class/Section _____

1. Describe three characteristics that differentiate a benign tumor from a malignant tumor.

2. What is the "multiple hit hypothesis" (as it relates to cancer)?

3. What is the role of the P53 gene in carcinogenesis?

4. What is the difference between carcinoma and sarcoma?

5. Besides surgery and chemotherapy, what are some nontraditional cancer therapy strategies that the mesothelioma patient has included in his daily regimen? Do you think any of these approaches are prolonging his life?

19

PLANT BIOTECHNOLOGY

INTRODUCTION

The term *plant biotechnology* describes the application of plant biology. Plant biotechnologists employ recombinant DNA technology to introduce genes encoding desirable characteristics into target plants by various methods of DNA transfer. The methods of DNA transfer vary widely. DNA can be physically introduced into plants using particle bombardment or electroporation. Genetic material may also be introduced biologically by using a bacterial vector (*Agrobacterium*). Target tissue for plant genetic engineering is often either rapidly growing embryogenic tissue or tissue that will ultimately give rise to shoots.

The first genetically engineered plants were produced in 1983. Since then several commercial applications of plant biotechnology have made their way into the public market. Genetically engineered tomato plants are available for purchase in grocery stores across the country. These tomatoes contain a modified fruit ripening gene that allows the fruit to remain firm long after picking leading to increased shelf life and enhanced flavor.

There are some sectors of the public that view plant biotechnology with an ethical distaste. These individuals feel that it is inappropriate to "mess" with nature, and fear a release of an "Andromeda strain." Proponents of the technology promote a future that will utilize less pesticides and herbicides and produce greater yields for farmers. For example, insect-resistance corn, cotton, soybean, and tomato plants containing the Bt gene (from *Bacillus thurengiensis*) can produce their own natural insecticide, thus relying less on chemical application.

In addition to herbicide and pesticide resistance, viral resistance is also a desirable characteristic of plants. Introduction of a specific portion of the viral genome into plants via genetic engineering has created several virus-resistant crop plants. Viral-resistant potato, squash, papaya, and tomato have now been obtained in the laboratory and evaluated in the field.

Considering the previously mentioned benefits of plant biotechnology, genetically engineered plants may soon become the norm rather than the exception on your table.

CHECK FOR URL UPDATES

Before beginning the activities in this chapter, check the Wadsworth Biology Resource Center home page for URL updates and information:

http://www.wadsworth.com/biology

Once you have arrived at the Wadsworth Biology Resource Center home page, select General Biology and locate this book. Any updates should appear.

ACTIVITY 1

In this activity you will visit the Center for Soybean Improvement site, which contains information from researchers at The Ohio State University, the University of Georgia, and the University of Kentucky. This joint research project is funded by the United Soybean Board, which has established the center to promote research in genetic engineering systems for soybeans so that the process will become a routine and an affordable means for soybean improvement.

Soybeans are the second largest crop grown in the United States and are the number one protein and oil crop. In order to increase soybean yield, and lessen or eliminate pesticide application, genetic engineering of soybeans is a focus of not only universities but also the biotechnology industry. In this activity you will learn about some approaches and techniques used in the pursuit of genetically engineered crops.

PART 1

1. Using Netscape, or another Web browser, type in the following address at the URL prompt:

   http://mars.cropsoil.uga.edu/homesoybean/index.htm

2. Select Progress.

3. Read the information about the two accomplishments of the Center for Soybean Improvement.

4. Answer the questions regarding soybean with a synthetic CryIA(c) Bt gene on the Soybean Center Activity Worksheet.

5. Return to the Soybean Center home page by using the "Back" key of your Web browser.

PART 2

1. Select Protocols.

2. Select Regeneration via Somatic Embryogenesis.

3. Read the protocol, paying special attention to the photographs of the plant structures being used or produced.

4. Answer the questions regarding somatic embryogenesis on the Soybean Center Activity Worksheet.

ACTIVITY 2

Now that you have some feel for the type of research taking place in the field of genetic engineering, it's time to visit a site that provides an electronic forum for the public debate over biotechnology. At the BIOSIS site you will read an article on high-tech food. This paper will raise some thought provoking questions about safety, economic concerns, regulatory processes, and the ethical and moral issues associated with biotechnology. After you read the paper, you are invited to take part in the online survey about biotechnology.

PART 1

1. Using Netscape, or another Web browser, type in the following address at the URL prompt:

   http://www.scicomm.org.uk/biosis/index.html

2. Select Hi-tech Food.

3. Read the article on high-tech food by Dean Madden.

4. Answer the questions regarding high-tech food on the BIOSIS Activity Worksheet.

PART 2

1. Return to the BIOSIS home page by using your browser's "Back" key.

2. Select Participate in an online survey about biotechnology.

3. The survey will take from 15 to 20 minutes. In addition to asking your opinions, it will also ask you some factual questions that require a basic knowledge of genetics.

4. After you complete the survey, check to see how many of the questions on biotechnology you answered correctly.

Follow the directions from your instructor regarding the submission of your worksheets. Some instructors will want them e-mailed; others will want them handed in.

SOYBEAN CENTER ACTIVITY WORKSHEET

Name _____ Date _____

Class/Section _____

PART 1: PROGRESS
Soybean with a Synthetic CryIA(c) Bt Gene

1. What is the general function of the CryIA(c) Bt gene?

2. What DNA transformation technique was utilized to introduce this gene into the target tissue?

3. What is the target tissue?

Plant Biotechnology

PART 2: SOMATIC EMBRYOGENESIS PROTOCOL

1. What is the difference between a somatic embryo and a zygotic embryo? (You may have to use your textbook to help you with this distinction.)

2. What factors in this protocol appear to induce somatic embryogenesis? (*Hint:* Look at the composition of the media used throughout the experiment.)

BIOSIS ACTIVITY WORKSHEET

Name _____ Date _____

Class/Section _____

1. Besides tomato, describe four other crops that are targets of plant biotechnology.

2. Choose one of the environmental safety issues mentioned. In one paragraph describe your feelings regarding this issue.

3. Choose one of the social effects issues mentioned. In one paragraph describe your feelings regarding this issue.

4. Discuss the public health need for regulatory control and safety testing of these genetically engineered food products.

PART THREE

PRINCIPLES OF EVOLUTION

20

A TOUR THROUGH THE VIRTUAL PALEONTOLOGY MUSEUM

INTRODUCTION
Paleontology is the study of prehistoric life forms through evidence provided by fossil remains. By examining the Earth's geologic stratifications and the organization of fossils within the strata, paleontologists can place evolutionary events on the geologic time scale, thereby producing a history of the planet Earth.

CHECK FOR URL UPDATES
Before beginning the activities in this chapter, check the Wadsworth Biology Resource Center home page for URL updates and information:

http://www.wadsworth.com/biology

Once you have arrived at the Wadsworth Biology Resource Center home page, select <u>General Biology</u> and locate this book. Any updates should appear.

ACTIVITY

In this activity you will explore the University of California's Museum of Paleontology. This virtual site is organized like an actual museum with various wings and exhibits. Navigation through the site can be confusing, so before exploring, take note of the organization of the museum.

All exhibits in the museum are organized into three wings: Phylogeny, Geology, and Evolution. The Phylogeny wing features the diversity and relationships between major groups of taxon on Earth. The Geology wing explores the geologic history of the Earth and includes a discussion of both plants and animals and geological events during a particular time period. The Evolution wing provides information on some of the investigators and scientists who paved the way for and formulated modern theories of evolutionary biology. Both the Phylogeny wing and the Geology wing can be entered from the main entrance hall (first page) as well as from the other exhibits. The Evolution wing can only be entered from the main entrance.

Within each exhibit you can always get help from the Web Lift Bar found at the bottom of each page. By using the bar you can link to almost any other site within the museum.

You will begin this activity by going on a "guided" tour through the museum. Following the tour, explore the wings on your own.

PART 1

1. Using Netscape, or another Web browser, type the following address at the URL prompt:

   http://ucmp.berkeley.edu/axhibit/exhibits.html

2. Enter the Geology wing by taking the Web Lift to "Any Period."
3. Select the Proterozoic Period.
4. After reading the introduction, select Ancient Life.
5. Select Stromatolites and travel to the Phylogeny wing of the museum.
6. View the photographs of colonial and filamentous forms of cyanobacteria.
7. Select Life History and Ecology.
8. Read about present-day cyanobacteria and view the displays.

PART 2

1. Take the Web Lift to the Vendian period.
2. Read about the period.
3. View the exhibit.
4. Select Fossil Localities.
5. Select White Sea.
6. Recreate the White Sea exploration by selecting making camp, hiking up the cliff, or looking for fossils.

PART 3

1. Return to the museum's main entrance by retyping the site's URL.
2. Select the picture of Charles Darwin.
3. Select Systematics.
4. Learn something about Cladisitics by reading the information provided.

PART 4

This ends your "canned" tour of the museum. Now select a geologic period, and as time permits, explore, explore, explore! (Hint: the Hadean period will provide a very quick tour, the Cenozoic period tour could last for an hour or more. But remember, you can exit the site at any time.)

21

MUTATION

INTRODUCTION

Mutation is simply any heritable change in the genetic material of an organism. This change can occur in many ways including deletions, additions, or rearrangements of the genetic material. When the word mutation is first heard, a detrimental action often comes to mind, yet it is only through mutation that new forms of genes are made available as raw material for the process of evolution. Mutations occur at the level of the gene, or in the DNA. However, because the gene is ultimately translated into protein, the mutation is physically manifested at the level of the protein. As stated, mutation may be advantageous if it allows a species to ultimately adapt to a changing environment; however, mutation may also result in a non-functional protein ultimately leading to a genetic disorder.

In this activity you will use a mutation worksheet to review various types of mutations and their consequences. In order to complete this exercise you must have a genetic code table (codes and amino acids) available.

CHECK FOR URL UPDATES

Before beginning this activity, check on the Wadsworth Biology Resource Center homepage for URL updates and information:

http://www.wadsworth.com/biology

Once you have arrived at the Wadsworth Biology Resource Center, select General Biology and locate this book. Any updates should appear.

ACTIVITY

1. Using Netscape or another Web browser, type the following address at the URL prompt:

    http://www.gene.com/ae/AE/AEPC/WWC/1994/gene_action.html

2. Read the introduction to the exercise and familiarize yourself with the material mentioned in the section "Concepts/Key Terms".

2. Begin the exercise with question #7 in Part 2: gene mutations.

3. Answer questions 7 through 21 on the activity report sheet.

4. Scroll to part three of the site, "Chromosomal Rearrangements".

5. Use the diagrams labeled A through G to answer the questions on the activity report sheet.

MUTATION ACTIVITY WORKSHEET

Name _____ Date _____

Class/Section _____

Gene mutations

7.

8.

9.

10.

11.

12.

13.

14.

15.

16.

17.

18.

19.

20.

21.

Chromosomal Rearrangements

Which diagram illustrates the following chromosomal aberrations?

_____	Polyploidy
_____	Translocation
_____	Trisomy
_____	Deletion
_____	Monosomy
_____	Inversion
_____	Normal Condition

If you think any of the above conditions are not represented in the diagrams provided at the site, use the space below to illustrate your version of the abnormality.

22

JUMPING SPIDERS AND THE TREE OF LIFE

INTRODUCTION

A species is a group of individuals with a defined set of unique morphological characteristics that can successfully reproduce. Successful reproduction includes not only producing offspring, but producing offspring which are fertile. For example, the donkey and horse belong to two different species although phenotypically they are very similar and can be mated to produce offspring, a mule. The mule however is sterile and as such, the union is not considered successful.

Speciation, or the emergence of a new species occurs when a population of individuals from one species becomes reproductively isolated. This isolation frequently is due to geographical isolation resulting in reproductive isolation. Ultimately, the population that has been isolated will adapt to its new environment and form a new species.

It is possible to examine the emergence of new species as well as species relationships and common ancestors by constructing phylogenic trees. These trees show the degree of evolutionary relatedness between different species. The Tree of Life homepage is a place where one can access phylogenic trees for many species. When you browse through this site and are presented with a tree, there are several things you should remember: 1) the left side of the tree is the root, or common ancestor of all species, 2) time on each tree runs from left to right, with older events being on the left side of the tree, and 3) any extinct species is represented by a horizontal row of vertical lines.

Construction of the trees relies on organization into clades. Clades are groups of organisms which include the common ancestor of all members and all descendants of the most recent common ancestor. The word clade comes from the Greek *klados*, meaning branch or twig.

The genus you will investigating on the tree is *Habronattus*, a large group of ground dwelling jumping spiders, many of which are found in the desert southwest of the United States as well as in Florida. The *Habronattus* genus is represented by a large number of species and the tree is somewhat complex. Your objective will be to become familiar with tree organization and learn a little something about these interesting arachnids.

CHECK FOR URL UPDATES

Before beginning this activity, check on the Wadsworth Biology Resource Center homepage for URL updates and information:

http://www.wadsworth.com/biology

Once you have arrived at the Wadsworth Biology Resource Center homepage, select General Biology and locate this book. Any updates should appear.

ACTIVITY

1. Using Netscape or another Web browser, type the following address at the URL prompt:

   http://spiders.arizona.edu/salticidae/habronattus/habronattus.html

2. Print this page as well as the next page, so that you'll have a copy of the phylogenic tree. You will need this information to answer questions on the activity worksheet.

3. Select Introduction. Read this section. You will use this information to answer the questions on the activity worksheet.

4. Select forms and colors to view graphics of several species of *Habronattus*. Pay special attention to those characteristics used to separate the genus into various species.

5. Use the "Back" key of your browser to return to the previous page. Select courtship behaviors. View one or two Quicktime™ movies of the male spiders. (Caution these movies can take several minutes to download).

6. Use the "Back" key of your browser to return to the previous page.

7. Select Discussion of Phylogenic Relationships. You will use this information to answer the questions on the activity worksheet.

JUMPING SPIDERS ACTIVITY WORKSHEET

Name _____ Date _____

Class/Section _____

1. What two basic characteristics delineate the species of *Habronattus*?

2. Do you think it appropriate that this genus is speciated on the basis of male characteristics alone? Why or Why not?

3. Based upon your answer to question one, theorize as to how a new species of *Habronattus* might easily arise.

4. Using the phylogenic tree, determine if *H. moratus* is more closely related to *H. banski* or *H. orbus*? Defend your conclusion.

PART FOUR

EVOLUTION AND DIVERSITY

23

VIEWS OF THE OCEAN FLOOR AND PLATE TECTONICS

INTRODUCTION

The Earth's crust is divided into a number of plates that move, separate, and collide over periods of time. This movement and collision creates seismic activity resulting in earthquakes and volcanic eruptions. In addition, tectonic activity has for billions of years caused movement of the Earth's land masses. The movement of the continental land masses has contributed to some of the great extinctions of the geologic time scale as well as exerted an evolutionary pressure on species that have been forced to adapt to climatic changes caused by this continental movement. In addition, land bridges between continents formed and then disappeared allowing spreading as well as geographical isolation of many species.

The best place to study and observe plate tectonics is on the ocean floor. Until now we knew little about its topography. The ocean floor could eventually be mapped using modern high resolution sonar technology but that would take over one hundred years. It turns out that it is possible to map the sea floor by measuring minute variations in the ocean's surface. This data is then processed using computer algorithms to generate a map of the sea floor. The U.S. Navy's GEOSAT satellite and the European Space Agency's ERS-1 satellite can measure sea surface height to an accuracy of one inch. This information has now been declassified and is being used by researchers at the Scripps Institute of Oceanography in California to produce computer-generated maps of the ocean floor.

CHECK FOR URL UPDATES

Before beginning the activities in this chapter, check the Wadsworth Biology Resource Center home page for URL updates and information:

http://www.wadsworth.com/biology

Once you have arrived at the Wadsworth Biology Resource Center home page, select General Biology and locate this book. Any updates should appear.

ACTIVITY

In Part 1 of this activity you will visit the Discovery Channel's home page and access an article on visualization of the ocean floor through satellite and computer-generated information. In Part 2 you will see a computer-generated image of a recently discovered set of volcanoes on the ocean floor. In Part 3 you will learn how those seamounts were formed by the action of tectonic plates.

PART 1

1. Using Netscape, or another Web browser, type in the following address at the URL prompt:

   http://www.discovery.com/DCO/doc/1012/world/roundup032796/roundup.html

2. Select World, Original Stories.

3. Select W Science.

4. Select Previous Story Index.

5. Select Previous Web Round up.

6. Select 3.27.96 Satellites See More of the Sea Floor.

7. Read the introduction by M. Mitchell Waldrop.

PART 2

1. Type in the following address at the URL prompt:

   http://seawifs.gsfc.nasa.gov/OCEAN_PLANET/HTML/
   oceanography_recently_revealed3.html

2. Enlarge the graphic of the three seamounts in East Pacific Rise by selecting the small image.

PART 3

1. Go directly to the plate tectonics site at the following address:

   http://volcano.und.nodak.edu/vwdocs/vwlessons/plate_tectonics/introduction.html

2. Continue reading the information by selecting Next Section at the bottom of each page.
3. Read the information in the article and answer the questions on the Plate Tectonic Activity Worksheet at the end of this chapter.

Follow the directions from your instructor regarding the submission of your worksheets. Some instructors will want them e-mailed; others will want them handed in.

PLATE TECTONIC ACTIVITY WORKSHEET

Name _____ Date _____

Class/Section _____

1. List the three major layers of the earth.

2. What evidence did Alfred Wegener observe to support his theory of continental drift?

3. On which major tectonic plate is the United States?

4. What seismic activities define plate boundaries?

5. List three types of plate boundaries.

6. What type of plate boundary is the San Andreas fault?

24

EMERGING INFECTIOUS DISEASE

INTRODUCTION
In the late 1960s and early 1970s, many individuals in the United States felt that the battle against infectious disease had been won. Consequently, the nation's health care community turned its attention toward conquering cancer. However, since the mid-1970s, we have begun to take notice of new and emerging infectious diseases, including a resurgence of diseases we thought were under control, such as cholera, various food-borne infections, hepatitis, and group A Streptococci. In addition, since 1973, several major new etiological agents and their diseases have been identified, including AIDS, Lyme disease, Ehrlichiosis, Ebola hemorrhagic fever, and Hantavirus pulmonary syndrome. To add to the crisis, the war against still other infectious agents has become strategically complex as many organisms develop resistance to commonly used antibiotics.

The Centers for Disease Control and Prevention (CDC) is the federal government agency responsible for monitoring disease statistics and trends and also for recommending prevention guidelines. In this activity, you will access the CDC's home page. From this page, you can link to the CDC's weekly publication, Morbidity and Mortality Weekly Report, CDC Special Reports, and CDC Disease Guidelines. A hyperlink is also provided to the National Center for Infectious Disease's home page.

CHECK FOR URL UPDATES
Before beginning the activities in this chapter, check the Wadsworth Biology Resource Center home page for URL updates and information:

http://www.wadsworth.com/biology

Once you arrive at the Wadsworth Biology Resource Center home page, select General Biology and locate this book. Any updates should appear.

ACTIVITY

In this activity you will become acquainted with a new disease, Hantavirus pulmonary syndrome (HPS), and an old but reemerging disease, tuberculosis. You will use this newfound information to compare and contrast these diseases and perhaps formulate a strategy as to how the threat of emerging infectious disease should be handled by federal agencies and the nations' health care professionals.

PART 1

1. Using Netscape, or another Web browser, type the following address at the URL prompt:

    http://www.cdc.gov/

2. Select Health Information.

3. Select Tuberculosis.

4. Select TB Frequently Asked Questions.

5. Select all questions listed under "Introduction."

6. Either read the document while on line, download the file onto a disc, or print a copy of the article. You will need this information to answer several questions posed on the worksheets at the end of this chapter.

PART 2

1. After completing the above assignment return to the CDC home page (you can do this by either typing in the URL or using the "Back" button provided by Netscape or your Web browser).

2. Select Diseases.

3. Select Additional Disease Areas.

4. Select Hantavirus Pulmonary Syndrome.

5. Select Cause, Prevalence, and Prevention Information.

6. Either read the document while on line, down load the file onto a disc, or print a copy of the article. Again, you will need this information to answer several questions posed at the end of this exercise.

PART 3

Use the information you have read on TB and HPS to answer the questions about these two emerging infectious diseases on the Emerging Infectious Disease Worksheet.

PART 4

Want to learn more about the threat of emerging infections?

1. Return to the National Center for Infectious Disease home page by typing in the following URL:

   http://www.cdc.gov/ncidod/ncid.htm

2. Select Publications.

3. Select Brochures and Leaflets.

4. Select Emerging Infectious Disease Threats.

5. Download or print the brochure.

6. Formulate an opinion regarding the threat of infectious disease in the twenty-first century. Write out your response on the Threat of Infectious Diseases Worksheet.

Follow the directions from your instructor regarding the submission of your worksheets. Some instructors will want them e-mailed; others will want them handed in.

EMERGING INFECTIOUS DISEASE WORKSHEET

Name _____ Date _____

Class/Section _____

1. What type of infectious agent causes each disease (virus, bacterium, protista, fungi, etc.)?

2. Although both diseases are respiratory illnesses, they are transmitted in very different ways. Describe the way each infectious agent is spread.

3. What is the prescribed treatment or course of therapy for each disease?

4. What are the chances that the patient will recover?

5. What precautions should an individual take to lower the risk of contracting these illness?

THREAT OF INFECTIOUS DISEASES WORKSHEET

Name _____   Date _____

Class/Section _____

What steps do you think should be taken to combat th ethreat of infectious disease to the public health in the twenty-first century?

25

PROTIST SURVEY

INTRODUCTION
Long overlooked by many biologists, the protists (Kingdom, Protoctista) are among the most diverse and numerous living organisms on Earth. Many of these microscopic organisms are the foundations of both freshwater and marine ecosystems because photosynthetic forms serve as the producers in aquatic food chains. These simple eucaryotic cells also sit in an important evolutionary seat. Fossil evidence suggests that the protists are the transition group between their prokaryotic ancestors and multicellular descendants.

As previously stated, the protistans are a diverse group with both aquatic and terrestrial species, photosynthetic, heterotrophic, and saprophytic modes of food acquisition, free living and parasitic lifestyles, and sexual and asexual forms of reproduction.

In this activity, you will delve deeper into the protist kingdom by using the Protist Image Data homepage. Your objective will be to discover facts about a Protist genera not mention in your text.

CHECK FOR URL UPDATES
Before beginning this activity, check on the Wadsworth Biology Resource Center homepage for URL updates and information:

http://www.wadsworth.com/biology

Once you have arrived at the Wadsworth Biology Resource Center homepage, select General Biology and locate this book. Any updates should appear.

ACTIVITY

1. Using Netscape or another Web browser, type the following address at the URL prompt:

   http://megasun.bch.umontreal.ca/protists/gallery.html

2. Once at the protist image site, browse the images and select a protist of interest. Choose a genera not discussed in your text.

3. Select Appearance.

4. Print the image of your protist, then use the information provided in the text of the site to label the image as to size and internal as well as external structures.

5. Continue to work and read through the parts of the site (introduction, ultrastructure, life history, classification) which will provide you with information to answer the questions in the activity worksheet.

Attach your labeled image to the activity report sheet.

PROTIST SURVEY ACTIVITY WORKSHEET

Name _____ Date _____

Class/Section _____

1. List the genus and species of your protist, as well as common name.

2. What type of habitat(s) does this protist normally occupy?

3. Does this organism provide any significant evidence as to the evolution of multicellular eucaryotes?

4. How does this organism reproduce?

26

THE DIVERSITY OF LIFE: THE FUNGAL AND ANIMAL KINGDOMS

INTRODUCTION

The diversity of life on the planet Earth is staggering. From the smallest bacterium living in the human intestinal tract to the killer whale in its marine environment, living creatures occupy all sorts of different biomes, ecosystems, and niches. Unfortunately, the accelerated destruction of some of these ecosystems by human activities has forced us to look cautiously toward the future of certain species. Destruction of habitat has contributed to the reduction in population size or the extinction of many known plants and animals. Zoos have attempted to stem the rate of extinction through captive breeding programs with the resultant release of some species back into natural habitats. Through these efforts, many zoos have made significant contributions to the world wide conservation effort.

The Consortium of Aquariums, Universities, and Zoos (CAUZ) is a group that has formed to facilitate communications among scientists, educators, and zoo and aquarium personnel. It is hoped this exchange will contribute to the quantity and quality of research in zoos and lead to even greater success of captive breeding and release projects. In August 1995 the CAUZ site was established on the World Wide Web in order to provide an electronic forum through which scientists and aquarium and zoo personnel could network and share information to the benefit of all involved. This site provides links to over six hundred other Web sites, and is an invaluable tool when studying the diversity of life or health of an ecosystem.

In the following two activities you will use the CAUZ page to access the Tree of Life page from which you can learn almost anything about any living species. In Activity 1 you will explore the fungal phylum Ascomycota. In Activity 2 you will investigate beetles of the animal kingdom. In both activities, you will begin at the CAUZ home page.

CHECK FOR URL UPDATES

Before beginning the activities in this chapter, check the Wadsworth Biology Resource Center home page for URL updates and information:

http://www.wadsworth.com/biology

Once you have arrived at the Wadsworth Biology Resource Center home page, select General Biology and locate this book. Any updates should appear.

ACTIVITY 1

1. Using Netscape, or another Web browser, type the following address at the URL prompt:

    http://www.selu.com/~bio/cauz/

2. Select Explore the Web.

3. From Internet Links select Plants.

4. Select the Tree of Life.

5. From the Tree of Life page you can access a specific group of organisms in many different ways. After all, this is the Web! In the introduction to the Tree of Life page note the mention of beetles and sac fungi. These organisms are typed in blue, which indicates that we can link directly to the information about those two groups by simply clicking on the word or words in blue.

6. Select Sac Fungi.

7. While at this site, you will read the information on the Ascomycota (sac fungi), and then answer questions about this kingdom on the Fungal Kingdom Worksheet found at the end of this chapter.

ACTIVITY 2

1. Return to the Tree of Life home page by using the "Back" key of your browser.

2. Select Beetles in the introductory section of the page.

3. While on the beetle page remember that you can click on any photograph or highlighted word to hyperlink to another Web site. However, to answer the questions posed on the Beetle Activity Worksheet, simply remain within the document onscreen.

Follow the directions from your instructor regarding the submission of your worksheets. Some instructors will want them e-mailed; others will want them handed in.

116   Internet Activities for General Biology

FUNGAL KINGDOM WORKSHEET

Name _____   Date _____

Class/Section _____

1. List two parasitic or pathogenic members of the phyla other than the genera that cause Dutch Elm disease or Chestnut blight.

2. What are two chemical components of the cell walls of these fungi?

3. What symbiotic relationship do the Ascomycota and beetles share?

4. Describe two ways in which Ascospores may be dispersed.

5. Approximately how many years ago on the evolutionary timeline did Ascomycota first appear?

The Diversity of Life

6. Is there such a thing as male or female Ascomycota?

FUNGAL KINGDOM WORKSHEET

Name _____ Date _____

Class/Section _____

1. How many species of beetles have been described?

2. What is the approximate range in size of beetles?

3. How far back do beetle fossils date?

4. List three different ways beetles communicate.

5. Describe the two sets of wings on beetles.

27

BRYOPHYTA

INTRODUCTION

The plant kingdom consists of species that are photosynthetic and in general multicellular. The four major plant groups within the kingdom include the bryophytes, ferns, gymnosperms and angiosperms. These four groups demonstrate quite visually the evolutionary trends begun in the Devonian period of the Paleozoic era when the bryophytes first appeared and continue through the Cretaceous period of the Mesozoic when flowering plants appear. Within the kingdom, we see a decreased dependence on water for fertilization, development of vascular tissue for storage and transport of water and nutrients, less reliance on the gameophytic stage during the lifecycle and a general increase in plant size.

In this activity, you will survey the plant kingdom visually through the use of graphics and drawings. The activity will focus on the bryophytes, a small and often overlooked group containing mosses, liverworts, and club mosses. The bryophytes are of particular interest as they seem to represent an aquatic-to-terrestrial transition group that has moved onto land yet has no ability to store water, or fertilize in dry air and as such, must seek out environments where it can survive.

CHECK FOR URL UPDATES

Before beginning this activity, check on the Wadsworth Biology Resource Center homepage for URL updates and information:

http://www.wadsworth.com/biology

Once you have arrived at the Wadsworth Biology Resource Center homepage, select General Biology and locate this book. Any updates should appear.

ACTIVITY

1. Using Netscape or another Web browser, type the following address at the URL prompt:

   http://www.cc.manhattan.edu/science/biology/plants_new/intro/start.html

2. Select Click here for start.

3. Select Click here to begin.

4. Select the first green arrow to start at the beginning of the study.

5. Scroll down the page until you come to the plant kingdom. Select <u>4.0.3 Division bryophyta</u>.

6. Select <u>Characteristics of Division Bryophyta</u>. Read this information and use it to answer the questions on the activity worksheet.

7. Use the "back" key of your browser to return to the previous page then select <u>Order Sphagnales</u>. Read this information and use it to answer the questions on the activity worksheet.

8. Use the "back" key of your browser to return to the page listing all divisions of the plant kingdom, then select "Click <u>here</u> to connect to A List of the World's Most Endangered Bryophytes".

9. Read the information about several endangered species then answer the questions on the activity worksheet.

BRYOPHYTE ACTIVITY WORKSHEET

Name _____ Date _____

Class/Section _____

1. Describe four characteristics of brophytes that distinguish them from other plant groups.

2. How many species of *Sphagnales* are recognized. List the common name of one member of this order.

3. How do members of the order *Sphagnales* modify their environment to make it inhospitable for other organisms?

4. Why does decomposition occur very slowly in a peat bog?

5. List at least three general reasons why some species of bryophytes have become endangered.

28

THE BEAST OF BODMIN MOOR

INTRODUCTION
Evolutionary speaking, mammals are the most recent class of invertebrates to emerge. The group, with over 4500 species, demonstrates great diversity of body plan and ecological niche. Common to all mammals is an internal skeleton and a backbone atop which sits the skull which encasing the brain. Mammals also have a distinctive numbers, sizes, and types of teeth. Of particular interest to paleontologists is the unique shape and surface of the incisors, canines, and molars which provide clues as to what types of food and environments ancestral mammals experienced.

In this activity, you'll be visiting the Natural History Museum (United Kingdom) to work on the science casebook. Your goal will be to identify the mystery beast of Bodmin Moor. Skull shape and size as well as dentition will be some of the major clues you'll be using to solve the case.

CHECK FOR URL UPDATES
Before beginning this activity, check on the Wadsworth Biology Resource Center homepage for URL updates and information:

http://www.wadsworth.com/biology

Once you have arrived at the Wadsworth Biology Resource Center homepage, select General Biology and locate this book. Any updates should appear.

ACTIVITY 1

1. Using Netscape or another Web browser, type in the following address at the URL prompt:

   http://www.nhm.ac.uk/sc/

2. Select The Beast of Bodmin Moor

3. Read the information on each screen paying close attention to any clues that are presented regarding the beast.

4. After you have read the information select Continue to move on to the next screen. In addition to the Continue button, you also will find a Back button on the bottom of the page. Use that button anytime you need to review a clue.

5. A line scale is provided when you are looking at the skulls. Use this scale to make comparisons between the beast skull and known skulls from the museum's collection.

6. On some screens you may be required to select Yes or No or A or B. After making your selection, you will be presented with immediate feedback.

7. Continue through the screens until you have solved the mystery of the Beast of Bodmin Moor.

8. Answer the questions on the Beast of Bodmin Moor activity worksheet.

ACTIVITY 2

1. The site provides a feedback section onto which you can post messages about the case. If you found any step in the beast identification process difficult to understand or illogical, please post a message to the discussion board. Your message should be addressed within the week.

BEAST OF BODMIN MOOR ACTIVITY WORKSHEET

Name _____   Date _____

Class/Section _____

1. To which mammalian family does the beast belong? What clues led you and the museum scientist to draw this conclusion?

2. Why is this not the skull of a tiger, or a snow leopard?

3. What is the final identification of the Beast of Bodmin Moor?

4. What clues were provided that allowed you to conclude that the beast could not have died on Bodmin Moor?

The Beast of Bodmin Moor

29

PRIMATES AND HUMAN EVOLUTION

INTRODUCTION
The taxonomic order of the primates holds great fascination for most students because, after all, that is the place you will find *Homo sapiens*.

As in most taxonomic classification schemes, the group is very large and diverse, and evolutionary relationships among individuals within the group are always the subject of heated debate. In the next two activities you will explore the primates both visually, and in evolutionary terms.

CHECK FOR URL UPDATES
Before beginning the activities in this chapter, check the Wadsworth Biology Resource Center home page for URL updates and information:

http://www.wadsworth.com/biology

Once you have arrived at the Wadsworth Biology Resource Center home page, select General Biology and locate this book. Any updates should appear.

ACTIVITY 1

In this activity you will explore the visual diversity of the order primate through the use of the Primate Gallery. The Primate Gallery includes photos, illustrations, and paintings, some of which are available for use as public domain materials. The Primate Gallery Archive provides photographs of each family within the order. The gallery also provides vocalizations so that you can listen to primate calls in the wild. Because of the visual aspect of the site and the speed of your computer, it may take a minute or two to load these pages.

1. Using Netscape, or another Web browser, type in the following address at the URL prompt:

   http://www.selu.com/~bio/PrimateGallery/

2. Select Primate Gallery Archive.
3. View the photographs of all thirteen representative families.
4. Select Primate of the Week. Select the name of the primate to access details on that particular animal.
5. Answer questions concerning this primate on the Primate Activity Worksheet found at the end of this chapter.

As an option, and if you have the appropriate hardware, you can access the Primate Vocalizations to get a feel for the variety of auditory communication within the order.

ACTIVITY 2

Now let's look a little closer at the family Hominidae, in particular, the subfamily hominid. Evolutionary theory regarding the origins of humans has always been a controversial subject. Scientific theory and religious dogma seem to be at odds in any discussion of human evolution. Even among anthropologists (scientists who study human origin) the fossil evidence is frequently ambiguous and controversial. Also, the announcement of several recent finds as well as an unending redesignation of fossil genera make the subject somewhat difficult to present in an unambiguous manner. Ultimately the interpretation of the data and hominid evolutionary events becomes the responsibility of the reader. It is therefore to the reader's advantage to access as many different interpretations of the information as possible.

In this activity you will access the Primate Info Net maintained by the University of Wisconsin–Madison Regional Primate Research Center. From the Primate Info Net you will read an article by Jim Foley found in the Talk.Origins archive. Because this article is based on the recognized fossil evidence, it has been placed on the Primate Info Net as a source for the scientific explanation of human origins.

As hominoid fossils continue to be discovered, the human tree often changes. Since 1994 there have been several announcements of major fossil finds that shed new light on our early ancestors. The article you will access on the Primate Info Net is a complete summary of the fossil evidence to date. Included in the article are photographs and illustrations of several important fossils. And finally, the author of the article will broach a sensitive subject regarding how creationists reconcile the fossil evidence to their beliefs.

1. Return to Primate Net Sites by using the "Back" key. Or, you may begin the search by repeating the steps in Activity 1. Type in the following address at the URL prompt:

   http://www.primate.wisc.edu/pin/

2. Select Information Resources.

3. Select About Primates.

4. Select Evolution.

5. Select Fossil Hominids by Jim Foley.

6. Read all sections under Hominid Species and Hominid Fossils.

If you're curious about other points of view, you can access the Talk.Origins newsgroup by clicking on the "Return" area of the Talk.Origins Archive title banner at the top of the page of the Jim Foley article. Be prepared for some very heated discussion!

Follow the directions from your instructor regarding the submission of your worksheets. Some instructors will want them e-mailed; others will want them handed in.

PRIMATE ACTIVITY WORKSHEET

Name _____ Date _____

Class/Section _____

1. What is the family of the primate of the week?

2. List or describe three characteristics of this animal.

HUMAN EVOLUTION WORKSHEET

Name _____   Date _____

Class/Section _____

1. Describe two of the earliest species of hominid fossils that are not included in the discussion in your text book.

2. In which region of Africa did Donald Johanson make his discovery of the fossil Lucy?

3. To which genus and species have each of the following fossils been assigned?

    Nutcracker Man

    Taung Baby

    Lucy

PART FIVE

PLANT STRUCTURE AND FUNCTION

30

PLANT MOVEMENT

INTRODUCTION

Other then in growth or elongation processes, we would not think of plants as being able to move. However, if one has the proper detection system, the movement of plants can be seen. This movement is deliberate and induced by environmental stimuli such as gravity and sunlight. Movements such as nutation and sleep movements of plants have also been noted. It is not clear as to why plants make these movements, however the time lapse video of these activities makes for interesting viewing.

In this activity, you'll view actual time lapse movies of gravitropism, phototropism, nutation movement, sleep movement and also some virtual animations of the more common growth processes including root and shoot elongation. In order to view the movies or animations you must have QuickTime™ loaded. If you do not have QuickTime™ on your computer, the software can be easily downloaded at the site.

Check for URL Updates

Before beginning this activity, check on the Wadsworth Biology Resource Center homepage for URL updates and information:

http://www.wadsworth.com/biology

Once you have arrived at the Wadsworth Biology Resource Center homepage, select General Biology and locate this book. Any updates should appear.

ACTIVITY

1. Using Netscape or another Web browser, type the following address at the URL prompt:

    http://sunflower.bio.indiana.edu/~rhangart/plantmotion/PlantsInMotion.html

2. Scroll down the page and select GO TO THE MOVIES

3. Pull down the movie menu for arabidopsis movies (located in the frame on the left side of the screen).

4. Select and view the movies for Root growth, Phototropism, and Nutation.

5. Under the heading Other Movies, select Sleep movement.

    Note: these movies will take a while to download, particularly if you are using a modem.

6. Type in the following address at the URL prompt:

    http://www.ctpm.org/Programs/IPI/ipivp.html

7. Select Animations.

8. View the following virtual plant movies, (Cotton), Visualising internal state of plant parts, Schematic and realistic representation of shoot development, and (Bean) Development of root and shoot.

31

SEED DISPERSAL

INTRODUCTION

Gymnosperms and angiosperms are by far the most successful groups of plants. This success is owed to the evolution of the seed. The seed with its tough seed coat can survive harsh conditions including drought, hot or cold weather. Also, nutrients are stored within the seed and are made available to the germinating embryo before the plant can fend for itself.

The seed may be dispersed in a variety of ways. Some seeds are dispersed by the wind, others produce hooks or burrs which will allow them to hitch a ride on the fur or feathers of animals, some seeds are surrounded by sweet fruits which are consumed and then excreted along with feces, and finally other seeds are dispersed by water currents. The structure of the seed often provides telltale clues as to how seed dispersal may occur. Even modern day forensics have taken advantage of the uniqueness of seeds and their dispersal patterns by using seeds found on or near a corpse to determine if the body has been moved, or to trace and individual's movement in a rural setting before death.

The Seeds of Life homepage presents the Mystery Seed Contest. Not only will you learn about seeds but you will also get an opportunity to win a contest.

Check for URL Updates

Before beginning this activity, check on the Wadsworth Biology Resource Center homepage for URL updates and information:

http://www.wadsworth.com/biology

Once you have arrived at the Wadsworth Biology Resource Center homepage, select General Biology and locate this book. Any updates should appear.

ACTIVITY

1. Using Netscape or another Web browser, type the following address at the URL prompt:

   http://versicolores.ca/SeedsOfLife/home.html

2. Select What is this Seed?

3. View each mystery seed in turn by clicking on the graphic to enlarge the photograph.

4. Answer the questions about these seeds on the activity worksheet. If you are familiar with this seed and have a clue as to its identity you can post that information on the web page.

5. This site provides a nice review on different fruit types. To access this information from the homepage, select Fruits and Seeds.

SEED DISPERSAL ACTIVITY WORKSHEET

Name _____ Date _____

Class/Section _____

1. Describe or draw the unique physical characteristics of each seed.

   Seed 1

   Seed 2

   Seed 3

2. Given the seed's outward appearance, what would be the most logical means of dispersal for each seed?

   Seed 1

   Seed 2

   Seed 3

3. List or describe any common plants that you are familiar with that produce seeds which resemble the mystery seed.

   Seed 1

   Seed 2

   Seed 3

32

SECONDARY PLANT PRODUCTS AND POISONOUS PLANTS

INTRODUCTION

Plants provide us with food, fiber, building materials, and oxygen. Without them we would cease to exist. Plants also produce important secondary compounds not integral to metabolism, yet necessary to protect the plant from predators, parasites, and infectious agents. Examples of secondary compounds include tannins, terpenes, terpenoids, phenols, and cyanogenic glycosides.

We exploit many secondary compounds in the field of medicine. Some medicinal secondary plant compounds include the following: salicylic acid, digitalis, taxol, opiates, caffeine, and quinine. There are certainly many additional secondary compounds in herbs and other plants that have yet to be discovered. Conversely, we are also aware that some secondary compounds are toxic or poisonous to humans and should therefore be avoided.

Poisonous substances are found in house plants, garden plants, and ornamentals, It is important to familiarize yourself with some of the more common poisonous plants. You might be surprised to find that many household plants are toxic and that some common garden plants have parts that are completely edible whereas other parts are poisonous.

CHECK FOR URL UPDATES

Before beginning the activities in this chapter, check the Wadsworth Biology Resource Center home page for URL updates and information:

http://www.wadsworth.com/biology

Once you have arrived at the Wadsworth Biology Resource Center home page, select General Biology and locate this book. Any updates should appear.

ACTIVITY

The U.S. Army Center for Health Promotion and Preventative Medicine maintains an Internet resource that is a guide to poisonous and toxic plants. The contents of the guide are divided into (1) house plants, (2) garden plants, (3) ornamental plants, and (4) an index containing wild plants. Information in each section includes the plants' scientific name, its common synonyms, a listing of the toxic portion(s) of the plant, and the major symptoms experienced on contact or ingestion.

In this activity you will search for toxicity information on commonly encountered house, garden, and ornamental plants.

PART 1

1. Using Netscape, or another Web browser, type in the following address at the URL prompt:

   http://chipmunk.apgea.army.mil/ento/PLANT.htm

2. Select Houseplants.
3. Scroll down the page until you come to *Diffenbachia*.
4. Use the Secondary Plant Compounds and Poisonous Plant Worksheet to record information about the characteristics of this plant.

PART 2

1. Use the "Back" key of your Web browser to return to the introductory page of this site.
2. Select Garden Plants.
3. Use the Secondary Plant Compounds and Poisonous Plant Worksheet to record information on the following three plants:
   Colchicum
   Lathyrus
   Rheum rhabarbarum

PART 3

1. Use the "Back" key of your Web browser to return to the introductory page of this site.
2. Select Ornamental Plants.
3. Use the Secondary Plant Compounds and Poisonous Plant Worksheet to record information about *Nerium oleander,* a poisonous ornamental plant.

Follow the directions from your instructor regarding the submission of your worksheets. Some instructors will want them e-mailed; others will want them handed in.

SECONDARY PLANT COMPOUNDS AND POISONOUS PLANT WORKSHEET

Name _____ Date _____

Class/Section _____

POISONOUS HOUSE PLANTS
Diffenbachia

1. What is (are) the common name(s) for this plant?

2. List the toxic part(s) of this plant.

3. What are the symptoms as a result of contact or ingestion of this plant?

POISONOUS GARDEN PLANTS
Colchicum

1. What is (are) the common name(s) for this plant?

2. List the toxic part(s) of this plant.

3. What are the symptoms as a result of contact or ingestion of this plant?

Lathyrus
1. What is (are) the common name(s) for this plant?

2. List the toxic part(s) of this plant.

3. What are the symptoms as a result of contact or ingestion of this plant?

Rheum rhabarbarum
1. What is (are) the common name(s) for this plant?

2. List the toxic part(s) of this plant.

3. What are the symptoms as a result of contact or ingestion of this plant?

POISONOUS ORNAMENTAL PLANTS
Nerium oleander
1. What is (are) the common name(s) for this plant?

2. List the toxic part(s) of this plant.

3. What are the symptoms as a result of contact or ingestion of this plant?

PART SIX

ANIMAL STRUCTURE AND FUNCTION

33

VIRTUAL FROG DISSECTION

INTRODUCTION
Traditionally, dissection has been used in introductory biology laboratories to study the anatomy and systems of representative species. Frogs, fetal pigs, cats, and sharks are some of the most commonly used animals. Although laboratory dissections are an invaluable tool when studying anatomy, many schools have moved away from their use. Cost, disposal, and student objection to dissection are but a few of the reasons that these laboratory exercises are becoming more scarce.

It is now possible to do virtual dissection from a site on the Web. Although some would agree that virtual dissection is not as effective a teaching tool as actual dissection, it does provide the student with some insight into animal anatomy without the effort, odor, and squeamishness that may accompany actual dissection. In addition, there is no cost or cleanup involved.

CHECK FOR URL UPDATES
Before beginning the activities in this chapter, check the Wadsworth Biology Resource Center home page for URL updates and information:

http://www.wadsworth.com/biology

Once you have arrived at the Wadsworth Biology Resource Center home page, select General Biology and locate this book. Any updates should appear.

ACTIVITY

In this activity you will access the virtual frog dissection site and carry out the dissection. There will be an icon available to click on in each part of the dissection so that you can practice or repeat what you've learned at any time. The four parts of the dissection include the following:

1. Preparation and setup for dissection
2. Skin incisions
3. Muscle incisions
4. Internal organs

There will be four layers of organs for you to view within the internal organ section.

As with any site that includes photographs and graphics, it may take a minute or so to load a section of the site. Please be patient.

If you have the ability to configure a video viewer on your computer, the site contains several quicktime movies within the various sections. Following completion of the activity you should be familiar with dissection procedures and be able to identify the major organs within the body cavity of the frog.

1. Using Netscape, or another Web browser, type in the following address at the URL prompt:

   http://curry.edschool.Virginia.EDU/go/frog/

2. Select Menu.

3. Begin the laboratory by reading the Introduction.

4. When you're ready to begin dissecting, select Preparation and Setup for Dissection. The procedure is self-explanatory and fairly intuitive. Follow the directions until you have completed the dissection.

34

HUMAN CROSS-SECTIONAL ANATOMY

INTRODUCTION

The Visible Human Project began in 1986 with a look toward the future of electronic communication and the possibility of creating an electronic image bank for the National Library of Medicine (NLM). In 1989 plans were finalized for the NLM to undertake a project to build a digital image library of a normal human cadaver. This image library was to include cryopreserved cross sections as well as digital images produced by computerized tomography (CT) and magnetic resonance imaging (MRI). In 1991 a contract was awarded to Victor M. Spitzer, Ph.D, and David G. Whitlock, MD, Ph.D, at the University of Colorado at Denver to produce the required images. After a two-year wait, the project located an appropriate male cadaver and began sectioning. MRI scans were taken at 4mm intervals and CT scans and cross sections were done at 1mm intervals. There are 1,871 sections for each mode. The images of these sections make up a data set that is 15 gigabytes in size! A data set for a female cadaver is also being produced. In the female the axial images will be at 0.33mm intervals, which will result in over 5,000 images producing a data set approximately 40 gigabytes in size!

Although a complete data set is available from the Visible Human Project, you must have a license agreement (which can be purchased) to access the total complement of images. However, samples of the images can be found at other sites around the Web. The Lumen page of Loyola University's Stritch School of Medicine is such a site. This site makes use of the Visible Human cross sections, as well as MRIs and CT scans. The site is designed to be used as part of the Structure of the Human Body curriculum at the Stritch School of Medicine. Initially the student selects a particular portion of the body (head, thorax, abdomen, etc.). The next page includes a visual orientation of the cross section on a live model. Then the student can select the type of image they would like to view. You can chose from either a labeled cross section, an unlabeled cross section, an MRI scan, a CT scan, or a video of the section.

CHECK FOR URL UPDATES

Before beginning the activities in this chapter, check the Wadsworth Biology Resource Center home page for URL updates and information:

http://www.wadsworth.com/biology

Once you have arrived at the Wadsworth Biology Resource Center home page, select General Biology and locate this book. Any updates should appear.

ACTIVITY

In this activity you will access the Lumen page, select a section of the body, view several images, and identify structures. Be patient—these can be very large files that take a few minutes to load. Also remember that the images you are viewing will be cross sections only 1mm thick, so it can be very difficult to identify common organs.

PART 1

1. Using Netscape, or another Web browser, type in the following address at the URL prompt:

    http://www.meddean.luc.edu/lumen/MedEd/GrossAnatomy/cross_section /index.html

2. Select the thorax region of the body by clicking on the appropriate label.
3. Select an unlabeled image in the thorax. Choose the fourth section from the top. Can you find the trachea on the image?
4. Return to the previous page by using the "Back" button of your browser.
5. Select the labeled image. Did you correctly identify the trachea?

PART 2

1. Return to the drawing of the human body by selecting "Back" twice.
2. Select the abdomen.
3. Select the unlabeled cross section at the very bottom of the model.
4. Can you locate the following three structures: colon, jejunum, and internal oblique muscle?
5. Return to the previous page by selecting the "Back" button.
6. Select the labeled cross section at the bottom of the model. Did you identify the colon, jejunum, and internal oblique muscle correctly?

PART 3

1. Return to the drawing of the human body by selecting the "Back" button twice.
2. Select another section of the human body.
3. Select any unlabeled cross section. Can you identify any structures or systems?
4. Return to the previous page and select the corresponding labeled image. Check to see if you were correct on your identification.

PART 4

Turn to the Human Cross-sectional Anatomy Worksheet at the end of this chapter and answer the questions regarding this site.

Follow the directions from your instructor regarding the submission of your worksheets. Some instructors will want them e-mailed; others will want them handed in.

HUMAN CROSS-SECTIONAL ANATOMY WORKSHEET

Name _____ Date _____

Class/Section _____

1. How valuable do you think the visible human site will be to medical students?

2. Do you think this project (VHP) or Web sites that use this data should replace hands-on autopsies by students with actual human cadavers?

3. Is this site useful to undergraduates for the study of human anatomy? Why or why not?

35

SCIENTIFIC JOURNALS

INTRODUCTION

Professional scientific societies establish journal publications in which researchers publish results of their experiments within their specialized discipline. Before publication, these articles are peer reviewed to make sure that the experiments are well designed and thorough and that the results are valid and reproducible. Once scientific studies are published, the research may provide a springboard for additional studies or it may provide a product for industrial or clinical use.

Electronic journals are now available on the World Wide Web. Material that was previously accessible only at large university libraries can now be accessed by all who have the computer hardware and software available. Unfortunately some journals only provide the article abstract on their homepage. Nevertheless, scientific information that was very hard to get in the past is easily communicated to a much larger audience.

In this activity you'll browse the Journal *Neuron* published by the Neurological Society of America. Your objective will be to find some recently published information describing some aspect of neuron function.

CHECK FOR URL UPDATES

Before beginning this activity, check on the Wadsworth Biology Resource Center homepage for URL updates and information:

http://www.wadsworth.com/biology

Once you have arrived at the Wadsworth Biology Resource Center homepage, select General Biology and locate this book. Any updates should appear.

ACTIVITY

1. Using Netscape or another Web browser, type the following address at the URL prompt:

   http://neuron.org

2. Browse the table of contents of the most recent edition of *Neuron*.

3. Select an article of interest to you. Read the article and be prepared to discuss it in class.

36

ALZHEIMER'S DISEASE

INTRODUCTION

Alzheimer's disease is a progressive, debilitating, neurological disease that was first described by Dr. Alois Alzheimer in 1906. The disorder most often occurs as the sporadic form which affects individuals over the age of 65. Less common is a familial form of Alzheimer's, that can have its onset as early as the fourth decade of life.

Alzheimer's disease is a type of dementia characterized by impaired memory, poor judgment, and personality changes that may be accompanied by inappropriate behavior. Diagnosis is made on the basis of personal history including a documentation of signs and symptoms over a period of time. CT (computerized tomography) scans may assist in making a probable diagnosis of Alzheimer's disease. Unfortunately a definitive diagnosis can only be made during autopsy when typical plaques and tangles can be observed in sections of brain tissue.

There is no cure for this disease, although some of the patient's symptoms can be treated (depression, nutritional inadequacies) with medication. As the disease progresses, the patient eventually becomes unable to take care of him or herself. The disease not only eventually kills the affected individual but also usually extracts an emotional toll on the family.

In this activity, you will view two sites which address Alzheimer's disease. The first site will provide information from the point of view of the clinician, while the second site will be maintained by the Alzheimer's Association, a patient support group.

CHECK FOR URL UPDATES

Before beginning this activity, check on the Wadsworth Biology Resource Center homepage for URL updates and information:

http://www.wadsworth.com/biology

Once you have arrived at the Wadsworth Biology Resource Center homepage, select General Biology and locate this book. Any updates should appear.

ACTIVITY 1

1. Using Netscape or another Web browser, type the following address at the URL prompt:

    http://werple.mira.net.au/~dhs/q&a.html

2. Select What is the cause of Alzheimer's disease?. Read the information presented. You may choose to print this page as it may assist you in answering the questions on the activity worksheet. Use the back key of your browser to return to the previous page.

3. Select Is there a treatment for Alzheimer's disease?. View the material as in step three.

4. Select How is Alzheimer's disease diagnosed? View the material as in step three.

5. Select Magnetic Resonance Imaging of an Alzheimer Patient. Select plaque and tangle to view the stained sections of an Alzheimer's brain. You may wish to print these pictures or upload them to a disc to help you answer the questions on the activity report sheet.

6. Select Tour 1 to view MRI images of an Alzheimer's patient.

7. Select normal central sulcus to view the same section in a normal patient.

8. As an option you can select "The Whole Brain Atlas" Home Page to view CT and MRI images of various parts of normal and diseased brain.

ACTIVITY 2

1. Using Netscape or another Web browser, type the following address at the URL prompt:

    http://www.alz.org

2. Select Caregiver Resources

3. Select fact-sheet on Alzheimer's disease. Use this information to answer the questions on the activity worksheet. Use the back key of your browser to return to the previous page.

4. Select In the News. Select one of the press releases. Read the information and note any progress that could provide an optimistic outlook for the Alzheimer's patient.

ALZHEIMER'S DISEASE ACTIVITY WORKSHEET

Name _____ Date _____

Class/Section _____

ACTIVITY 1

1. Describe the protein that has been linked to Alzheimer's disease.

2. Which chromosome(s) has (have) been implicated in late onset Alzheimer's disease? Which have been associated with early onset Alzheimer's?

3. Describe the microscopic appearance of an Alzheimer brain plaque. What normal brain structures make up the plaque?

4. Describe the appearance of an Alzheimer brain tangle. What structures form the tangle?

5. Describe the difference between the appearance of the central sulcus in the brain of the Alzheimer's patient and the brain of the normal patient.

ACTIVITY 2

1. Provide information on one report of Alzheimer's in the news.

37

COW'S EYE DISSECTION

INTRODUCTION

The eyes are our windows on the world, or so the saying goes. These sensory organs provide you with a wealth of information on the environment surrounding you. The eyes not only allow you to differentiate light and dark but also enable you to detect the subtlest differences in color, hue, and intensity. While shouldering an enormous sensory responsibility, the eye is actually a fairly simply structure composed of three basic layers. The outer layer consists of the sclera and cornea, the middle layer contains the iris, pupil, lens, aqueous humor, and vitreous body, and the inner layer contains the retina and part of the optic nerve.

The vertebrate eye is readily available and easy to dissect although some institutions may not have the facilities available to support such investigations. The Science Learning Network Homepage of the Exploratorium offers a cows eye dissection on-line. While not meant to take the place of actual dissection, this site offers a virtual dissection experience to those who may not have the opportunity otherwise. Those who are fortunate to be doing actual dissection can use the site as an on-line step by step guide.

Before beginning this exercise you may want to review the "Cows Eye Primer" to (re) acquaint yourself with the various parts of the eye. The site also provides video of the dissection procedure which you may access if you have the supporting hardware and software.

CHECK FOR URL UPDATES

Before beginning this activity, check on the Wadsworth Biology Resource Center homepage for URL updates and information:

http://www.wadsworth.com/biology

Once you have arrived at the Wadsworth publishing homepage, select General Biology and locate this book. Any updates should appear.

ACTIVITY

1. Using Netscape, or another Web browser, type the following address at the URL prompt:

   http://www.exploratorium.edu/learning_studio/cow_eye/index.html

2. Select <u>Step-By-Step: Dissecting a Cow's Eye</u>.

3. After completing each step, select the scalpel on the right side to continue. If you wish to return to the previous step, select the scalpel on the left.

4. As you dissect out each part of the eye, record your observations (relative location, size, shape, texture, color, etc.) and answer any questions on the activity report sheet at the end of this exercise.

5. There are 13 steps in this dissection. If at any time you do not know the meaning of a word that is highlighted in blue, select that term to go to the glossary for a definition.

6. If at any time you would to hear an explanation of the dissection step, select <u>RealAudio.</u>

COW'S EYE DISSECTION ACTIVITY WORKSHEET

Name _____ Date _____

Class/Section _____

1. Draw and/or describe the following parts of the eye. Include the function of each part.

   Sclera

   Cornea

   Aqueous Humor

   Lens

   Iris

   Vitreous humor

   Retina

(Question 1 continued)

Tapetum

Optic nerve

2. How was your vision altered as you looked through the cow's eye lens?

38

DIABETES

INTRODUCTION
The hormone producing cells of the pancreas carry out a delicate balancing act of maintaining appropriate glucose levels in the blood. Alpha cells of the pancreatic islets produce glucogon to raise blood glucose levels, beta cells produce insulin to lower blood glucose levels, and delta cells produce somatostatin which can block production of both glucogon and insulin.

In diabetes, little or no insulin is produced resulting in clinical disease. Untreated, diabetes can ultimately lead to blindness, kidney failure, stroke, heart disease and death. In the United States diabetes is the fourth leading cause of death by disease. It is estimated that 16 million people in the US have diabetes, yet half of them are not aware that they have the disease. Diabetes can be successfully controlled through diet, exercise, and if necessary, insulin injections.

There are many sites on the world wide web which address the subject of diabetes. It is critical that users evaluate these sites for dependability before using any information found on the site. In this activity, we will use reliable information found at the National Institute of Diabetes and Digestive and Kidney Diseases of the National Institutes of Health homepage and information from the American Diabetes Association homepage.

CHECK FOR URL UPDATES
Before beginning this activity, check on the Wadsworth Biology Resource Center homepage for URL updates and information:

http://www.wadsworth.com/biology

Once you have arrived at the Wadsworth Biology Resource Center homepage, select General Biology and locate this book. Any updates should appear.

ACTIVITY 1

1. Using Netscape or another Web browser, type the following address at the URL prompt:

   http://www.niddk.nih.gov/NIDDK_HomePage.html

2. Select Diabetes

3. Select <u>Diabetes Overview</u>.

4. Read the information. Answer the question about diabetes, types I and II, on the activity worksheet.

5. Using Netscape or another Web browser, type the following address at the URL prompt:

   http://www.diabetes.org/default.htm

6. Select <u>Diabetes Info</u>

7. Select <u>Take the Diabetes Risk Test</u>

8. Enter all information in the risk assessment, then submit your answers by selecting the Submit button.

9. Answer the questions on the activity worksheet concerning the risk assessment. DO NOT provide information regarding your assessment outcome. You assessment is confidential and should be used in a way that only you can deem appropriate.

10. Use the "back" key of your browser to return to the ADA homepage.

11. Select <u>Internet Resources</u>.

12. Select <u>Children with Diabetes</u>.

13. Select <u>Diabetes Basics</u> under the heading Diabetes Clinic.

14. Select <u>Humalog</u>™ (under the heading What is Insulin) to learn about a new genetically engineered form of insulin. Use this information to answer questions on the activity worksheet.

ACTIVITY 2 (OPTIONAL)

There are many non scientific sites that use humor, comics, and fun to teach something about diabetes. Listed below are two sites you may want to visit that not only teach but also entertain.

Captain Courage: **http://www.thehumanelement.com/courage**

The Virtual Diabetes Patient game: **http://www.nd.edu/~hhowisen/diabetes.html**

DIABETES ACTIVITY WORKSHEET

Name _____   Date _____

Class/Section _____

1. List three major differences between Diabetes Mellitus type I and type II.

2. List five warning signs of diabetes.

3. Why would having a baby over nine pounds (a question on the risk assessment) be considered a risk factor for diabetes?

4. How is Humalog™ biochemically different from regular insulin? How does this change make Humalog™ more effective than previously prescribed insulin?

5. Under what circumstances should Humalog™ not be prescribed?

39

SKIN CANCER

INTRODUCTION

The skin is the largest organ of the body. It is a hard, acidic, multilayered barrier that functions as the body's first line of defense against invading microorganisms. In response to ultraviolet exposure, the melanocytes in the epidermis of the skin produce melanin, a pigment which darkens the skin. In the early 20th century, people covered up their skin and avoided exposure to the tanning rays of the sun. White, creamy skin indicated a wealthy individual who did not have to work out of doors and be exposed to the sun. The social stigma of tanning changed dramatically in the latter part of the 20th century when tanning came to mean that an individual had plenty of leisure time and could enjoy the out of doors. Tanning came to signify a leisurely, healthy life style. With the skyrocketing incidence of skin cancer, we now realize that tanning is not the healthy activity a generation past believed it to be. Ultraviolet rays damage the genetic material of the cells. This damage may be repaired, but continued exposure to the sun over an extended period of time can lead to irreparable damage, transformed cells and cancer.

In this activity you will investigate the differences between melanoma and non-melanoma skin cancers.

CHECK FOR URL UPDATES

Before beginning this activity, check on the Wadsworth Biology Resource Center homepage for URL updates and information:

http://www.wadsworth.com/biology

Once you have arrived at the Wadsworth Biology Resource Center homepage, select General Biology and locate this book. Any updates should appear.

ACTIVITY

1. Using Netscape or another Web browser, type the following address at the URL prompt:

   http://www.noah.cuny.edu

2. Select Health Topics

3. Select Cancer

4. Scroll down the page until you come to Skin cancer. Select melanoma.

5. Select Description. Read the information and use it to answer the questions on the activity worksheet.

6. Use the "back" key of your browser to return to the previous page. Select Skin Cancer-Non melanoma.

7. Select Description. Read the information and use it to answer the questions on the activity worksheet.

SKIN CANCER ACTIVITY WORKSHEET

Name _____   Date _____

Class/Section _____

1. What type of cells have been transformed in melanoma?

2. List the warning signs of melanoma.

3. List the two types of non-melanoma skin cancers.

4. Compare possible outcomes or survival rates for melanoma and non-melanoma skin cancer.

40

HEART SOUNDS

INTRODUCTION
During the average life span the human heart beats over one billion times. The "beat" is a result of the four chambers of the heart going through phases of contraction and relaxation. When the atrioventricular valves close, the ventricles contract resulting in a "lub", and when the semiluminar valves close, the ventricles relax producing a "dup" sound.

As part of a routine physical examination, health care workers frequently listen to heart sounds expecting to hear the lub and dup of the normal heartbeat. On occasion, abnormal heart sounds may indicate some underlying cardiac conditions including arrhythmias and murmurs.

In this activity you will access the Heart Preview Gallery. You will listen to normal heart sounds, estimate heart rate, and then listen to a heart murmur. To listen to the heart sounds, your computer must be equipped with the appropriate hardware and software to support audio files. The heart sounds are presented at the site in three different formats. Choose that format supported by your computer.

CHECK FOR URL UPDATES
Before beginning this activity, check on the Wadsworth Biology Resource Center homepage for URL updates and information:

http://www.wadsworth.com/biology

Once you have arrived at the Wadsworth Biology Resource Center homepage, select General Biology and locate this book. Any updates should appear.

ACTIVITY

1. Using Netscape or another Web browser, type the following address at the URL prompt:

   http://www.fi.edu/biosci/preview/heartpreview.html

2. Select Hear.

3. Select Listen to the heartbeats.

4. On the next screen you will figure out which heart is beating at which speed.

5. To play the heart sounds, select the appropriate sound format for each example (1-4). When playing example one, select either .aa format, .aiff format or .wav format. Use your control panel to start the recording, listen to the heartbeat, and decide on the appropriate rate. Record your conclusions on the activity worksheet.

6. After you have listened to all four heart rates, use the back key of your browser to return to the previous page.

7. Select Heart Disease.

8. Select Heart Murmur.

9. Select Sound.

10. Listen to the heart sounds as in step 5. Describe the difference or the murmur you hear in this recording on the activity worksheet.

HEART SOUNDS ACTIVITY WORKSHEET

Name _____   Date _____

Class/Section _____

1. Record your estimates of heart rate here (four rate choices are listed at the site).

   Heart 1   _____

   Heart 2   _____

   Heart 3   _____

   Heart 4   _____

2. Describe the differences you hear in the heart sounds of a murmur.

3. Using what you know about the architecture of the heart, describe why you hear unusual sounds in a murmur or (in other words) what anatomical features of the heart could cause the sounds you hear? (You may need to consult your textbook for help.)

41

VIRTUAL IMMUNOLOGY LABORATORY

INTRODUCTION
In the specific immune response, antibodies are produced by plasma B cells in response to an antigen. This specific antibody then reacts with the target antigen to form a complex which may activate complement or induce phagocytic activity. Many diagnostic tests have been designed to exploit this very specific interaction of antibody and antigen. In this particular activity you will perform one of these tests, ELISA, in the virtual laboratory.

The Enzyme Linked Immunosorbent Assay detects either antigen or antibody associated with an illness or infectious agent. In order to visualize the antigen-antibody interaction, the test includes an enzyme which will produce a colored product when the correct substrate is added. In the ELISA you'll be performing, you will dilute out the patient's serum (containing antibody) and then calculate the titer of antibody (if present).

This virtual experiment requires you to perform each task. Tasks will be initiated by clicking with the mouse on various instruments and supplies. If any step is confusing or unclear, the laboratory notebook is available to provide you with an explanation of the protocol.

CHECK FOR URL UPDATES
Before beginning this activity, check on the Wadsworth Biology Resource Center homepage for URL updates and information:

http://www.wadsworth.com/biology

Once you have arrived at the Wadsworth Biology Resource Center, select General Biology and locate this book. Any updates should appear.

ACTIVITY

In order to view the animation and carry out the protocols, your computer must have the Shockwave™ plug-in. You can easily load Shockwave™ from the Virtual Laboratory site.

1. Using Netscape or another Web browser, type the following address at the URL prompt:

   http://www.hhmi.org/lectures/hiband/neat/start.htm

2. Select Find out more and test your diagnostic skills in our <u>Virtual Lab</u>!

3. Select [<u>CLICK HERE TO ENTER THE LAB</u>].

4. Read the laboratory protocol to determine your assignment.

5. Select (see <u>Figure1</u>) to see a schematic representation of the test. If possible print this graphic; you might use it later to answer the questions on the activity worksheet.

6. Use the "Back" key of your browser to return to the laboratory protocol.

7. At the top left of the screen under HHMI Virtual Lab sign, scroll down and then select <u>BACKGROUND</u>.

8. Read the section entitled, "How does this work?", to get an idea of what the experiment entails. You might need this information to answer the questions in the activity worksheet.

9. Use the "Back" key of your browser to return to the laboratory protocol.

10. Select <u>Click here to start</u> under the graphic on the right.

11. Follow the instructions under each graphic to carry out the protocol.

12. As you go through all eleven steps there will be a description of the lab protocol in the lab book on the left of the screen. In addition, the left screen will have a <u>Why?</u> selection. If you do not understand the protocol, select <u>Why?</u> for a detailed explanation. This information may also help you answer questions in the activity worksheet.

13. At the end of the experiment, calculate the titer of all three patients and make your diagnosis.

VIRTUAL IMMUNOLOGY LABORATORY ACTIVITY WORKSHEET

Name _____   Date _____

Class/Section _____

1. What disease are you attempting to diagnose?

2. Are you detecting antibody or antigen in this ELISA?

3. Why can this technique be described as a "sandwich" assay?

4. List the components of the positive control.

5. List the components of the negative control.

6. Calculate the titer for

    Patient A:

    Patient B:

    Patient C:

7. Did you carry out all protocol steps correctly? If not, what errors did you make?

8. Describe possible experimental results if you were to make the following errors in the protocol.

Failure to dilute patient's serum properly:

Failure to adequately wash the ELISA plate after adding the patient's serum and incubating at 37°C for 15 minutes.

Failure to add rabbit anti-human antibody to the ELISA plate:

Failure to add the buffered solution containing horseradish peroxidase to the ELISA plate:

42

ASTHMA

INTRODUCTION

Asthma is a chronic, obstructive lung disease characterized by a narrowing of the airway due to muscle constriction and increased mucus production in response to external stimuli. Allergies, stress, or exposure to environmental pollutants can all cause an acute asthma attach. These attacks can be life threatening; therefore, it is important that the disease be managed through the use of medication.

The University of Virginia Health Sciences Center Children's Medical Center provides an asthma tutorial for patients and their families on-line. This site provides animation, audio, and graphics to provide the lay person or patient with a good understanding of the disease.

CHECK FOR URL UPDATES

Before beginning this activity, check on the Wadsworth Biology Resource Center homepage for URL updates and information:

http://www.wadsworth.com/biology

Once you have arrived at the Wadsworth Biology Resource Center homepage, select General Biology and locate this book. Any updates should appear.

ACTIVITY

1. Using Netscape or another Web browser, type the following addresss at the URL prompt:

    http://galen.med.virginia.edu/~smb4v/tutorials/asthma/asthma1.html

2. Select <u>What is it?</u> Read the material, and use this information to answer questions on the activity worksheet.

3. Use the "Back" key of your browser to return to the tutorial introduction, and then select <u>Why Does it Happen</u>. Read the material, and use this information to answer questions on the activity worksheet.

4. Select the icon to listen to the recording of an asthmatic individual breathing. (Note: these sound files are not fully downloaded until the control bar appears in the browser frame.)

5. Select the icon to listen to the recording of a healthy individual breathing.

6. View the animation of normal breathing and then view the animation of breathing during an asthma attack.

7. Use the "Back" key of your browser to return to the Asthma title page.

8. Select <u>How Do We Treat It?</u>

9. Read this information and then answer the questions on the activity report sheet.

ASTHMA ACTIVITY WORKSHEET

Name _____ Date _____

Class/Section _____

1. Describe the appearance of the airways in the asthmatic patient during an asthmatic attack.

2. Describe the differences that you hear between the breathing of the normal patient and the asthmatic patient.

3. List the two types of medications prescribed to asthmatic patients. Describe the action of these medications.

43

THE FAST FOOD DIET AND NUTRITION

INTRODUCTION

Heterotrophic organisms do not have the ability to produce their own food and as such must derive energy and nutrients from preformed food sources. The largest heterotrophs—animals—rely on a digestive system to convert ingested food into nutrients and, ultimately, energy. In order to remain healthy, animals, and in particular humans, must eat a well-balanced diet containing varying amounts of carbohydrates, proteins, lipids, vitamins, and minerals.

In humans, diet has been shown to play a significant role in such diseases as diabetes, hypertension, cardiovascular illness, cancer, and tooth decay. Since the turn of the century, the fat intake of Americans has risen to approximately 40 percent of the total diet. Two federal agencies—the United States Department of Agriculture and the Department of Health and Human Services—have suggested that fat intake be reduced to less than 30 percent of the diet. In addition, cholesterol intake should be reduced by one half and sodium intake by two thirds. Unfortunately, in our fast-paced society many of us have turned to fast food restaurants for our meals. A reliance on these types of restaurants may lead many of us to consume even more fat, cholesterol, and sodium. That is certainly not to say that these establishments do not give you the option of purchasing healthy foods. It is important for the consumer to be aware of the nutritional content of the foods on the menu and to make the best selection based on nutritional value.

CHECK FOR URL UPDATES

Before beginning the activities in this chapter, check the Wadsworth Biology Resource Center home page for URL updates and information:

http://www.wadsworth.com/biology

Once you have arrived at the Wadsworth Biology Resource Center home page, select General Biology and locate this book. Any updates should appear.

ACTIVITY

The consumer division of the Minnesota Attorney General's Office has published a handbook of fast food facts. This information is available on the World Wide Web at the "Food Finder" site. In this activity you will access this information, gather data on several of your favorite fast foods, and design a fast food menu that is not only tasty but also healthful.

PART 1

1. Using Netscape, or another Web browser, type the following address at the URL prompt:

    http://www.olen.com/food/index.html

2. Search three restaurants for the healthiest hamburger choice.
    a. Select a restaurant name (e.g., McDonald's).
    b. Type in Hamburger.
    c. Press the "Fire up the Deep Fryer" button to search the index.

3. Record your data and answer the questions on the Fast Food and Nutrition Worksheet at the end of this chapter.

PART 2

1. Repeat the three steps in Part 1 to find information on the healthiest chicken choice.

2. Record your data and answer the questions on the Fast Food and Nutrition Worksheet.

PART 3

1. Use this site to design a meal containing three food choices—entree, side dish, and dessert—that would have a total of less than 15g of fat, 35mg of cholesterol, and 500mg of sodium.

2. Record your findings on the Fast Food and Nutrition Worksheet.

Follow the directions from your instructor regarding the submission of your worksheets. Some instructors will want them e-mailed; others will want them handed in.

FAST FOOD AND NUTRITION WORKSHEET

Name _____ Date _____

Class/Section _____

PART 1: BURGER SURVEY RESULTS

Restaurant	Fat (g)	Cholesterol (mg)	Sodium (mg)

Which restaurant serves the healthiest hamburger?

PART 2: CHICKEN SURVEY RESULTS

Restaurant	Fat (g)	Cholesterol (mg)	Sodium (mg)

Which restaurant offers the healthiest chicken choice?

PART 3: MENU

Food Item   Fat (g)   Cholesterol (mg)   Sodium (mg)

_____

_____

_____

Totals _____

44

KIDNEY DISEASE SEARCH

INTRODUCTION

The kidney is responsible for filtering metabolic wastes as well as maintaining electrolyte and solute balance in the body. The functional unit of the kidney is the nephron. Each kidney contains over two million nephrons and it is on the surface of the cells of the nephron that filtration, secretion and adsorption of materials takes place. Because the kidney is responsible for filtering out metabolic wastes it is environmentally assaulted on a regular basis and frequently will succumb to inflammation, infection, or structural abnormalities. Diseases of the kidney range from those that are very mild such as minor inflammation and infection to those that are very severe (polycystic kidney disease, end stage renal disease). Infections of the kidney are usually controlled with antibiotics, however severe forms of kidney disease may require dialysis and ultimately transplantation.

There are many sites on the World Wide Web that address the subject of kidney disease. Some of these sites are maintained by government agencies, while others have been developed by large medical centers. Still other sites may be maintained by individuals who have experienced kidney disease.

In this activity you will use the search engine of your choice to find information about the various types of kidney disease. Caution: when using a search engine, you are likely to encounter sites that provide unreliable information.

CHECK FOR URL UPDATES

Before beginning this activity, check on the Wadsworth Biology Resource Center homepage for URL updates and information:

http://www.wadsworth.com/biology

Once you have arrived at the Wadsworth Biology Resource Center homepage, select General Biology and locate this book. Any updates should appear.

ACTIVITY

1. Select the "Net Search" button on the main toolbar of your browser. Your browser will automatically load one of the search engines.

2. Once the browser has loaded a search engine, you have the option of selecting one of the other search engines which are listed on the page you are viewing. If you would prefer to use a search engine different than the one loaded by your browser, select it at this time.

3. Enter key words for your search. You can simply enter "Kidney Disease" or you can choose to enter the name of a specific kidney disease.

4. After a short period of time the search engine will return a list of several sites which correspond to your key words. Scroll down the list until you find a site that seems appropriate.

5. Select the site by clicking on the highlighted words.

6. Browse through the site. Keep in mind that you're looking for reliable information to supplement your text book on the subject of kidney disease.

45

THE VISIBLE EMBRYO: EMBRYONIC DEVELOPMENT FROM WEEK ONE THROUGH WEEK FOUR

INTRODUCTION

During the process of fertilization, the genetic components of egg and sperm come together to direct formation of a human zygote. From this union of two small cells, a complex, multicellular human being composed of trillions of cells is produced. During early development the cells undergo a very specific series of divisions, differentiation, and developmental processes to form tissues and ultimately organs and organ systems. This process of development in which a zygote becomes a fully formed human is accomplished by a truly remarkable series of events that are very visual and graphic in nature.

CHECK FOR URL UPDATES

Before beginning the activities in this chapter, check the Wadsworth Biology Resource Center home page for URL updates and information:

http://www.wadsworth.com/biology

Once you have arrived at the Wadsworth Biology Resource Center home page, select General Biology and locate this book. Any updates should appear.

ACTIVITY

The University of California at San Francisco has established a visible embryo site that presents the first four weeks of development through the use of photographs and graphics as well as text. Week one includes a description of fertilization and cleavage; week two focuses on implantation; week three illustrates the process of gastrulation; and during week four, longitudinal folding and the emergence of embryonic tissues is described. Following review of each week of developmental events, you can self-test with an online quiz to see if you've mastered the subject. Take note: The questions can be very difficult. Therefore, read each section carefully paying special attention to the terminology.

PART 1

1. Using Netscape, or another Web browser, type the following address at the URL prompt:

   http://visembryo.ucsf.edu/

2. Select <u>Fertilization</u>, or <u>Week One</u>.

3. Some images will provide more details if you select <u>Click here for more details</u>.

4. Continue reading the information.

5. After you have completed reviewing the first week of development, scroll up to the top of the page and select the <u>Self-Quiz</u> box. There are six multiple choice questions in this self-quiz.

6. Answer each question by selecting the correct answer and then selecting <u>Done</u>. Netscape, or your browser, will contact the host site to determine if you have answered the question correctly. Do not move on to the next question until you have discovered the correct answer to the previous question.

PART 2

1. Following completion of the first week's quiz, proceed to the information regarding the second week of development by selecting <u>Week Two</u> from the menu at the top of the page.

2. After reviewing the information, take the self-quiz. This quiz has only three questions.

PART 3

1. Following completion of the second week's quiz, proceed to the information regarding the third week of development by selecting <u>Week Three</u> from the menu at the top of the page.

2. After reviewing the information, take the self-quiz. The quiz contains five questions.

PART 4

Complete this activity by reviewing the information on <u>Week Four</u>, and then taking the quiz at the end of the section. This quiz has five questions.

46

SEXUALLY TRANSMITTED DISEASES

INTRODUCTION

Sexually transmitted disease (STD) prevention is of primary importance if we are to achieve the goals of improving women's and infants' health and to attain the national health status objectives for STDs (HP2000, Dept of Health and Human Services 1995). The good news is that the incidence of gonorrhea and syphilis are at their lowest levels since the early 1960's and 1970's respectively. Unfortunately, those public health gains are overshadowed by the increasing incidence of chlamydial as well as HIV infection. The highest rates of chlamydial infection are in adolescents, and in that group, female adolescents who are economically disadvantaged are particularly hard hit. Although syphilis and gonorrhea are decreasing in incidence, the emergence of antibiotic resistant strains of gonorrhea is particularly troublesome.

The Centers for Disease Control (CDC) is the governmental agency responsible for monitoring the incidence of disease in the United States. Each year, the CDC publishes a Sexually Transmitted Disease Surveillance report which includes reportable disease statistics collected from the previous year. This information is available on line at the CDCs homepage.

In this activity you will access the 1995 report (the 1996 report has not been published as of this date) and attempt to draw some conclusions regarding trends in bacterial STD incidence.

CHECK FOR URL UPDATES

Before beginning this activity, check on the Wadsworth Biology Resource Center homepage for URL updates and information:

http://www.wadsworth.com/biology

Once you have arrived at the Wadsworth Biology Resource Center homepage, select General Biology and locate this book. Any updates should appear.

ACTIVITY 1

1. Using Netscape or another Web browser, type the following address at the URL prompt:

   http://www.cdc.gov/

2. Select Health Information.

3. Select Sexually Transmitted Disease.

4. Under Other Resources, select 1995 Sexually Transmitted Disease Surveillance Report.

5. Select Summary Tables under the Detailed Tables heading.

6. Select Table 1.

7. Print the table. You will use this information to answer questions in the activity worksheet.

8. Select Table 2.

9. Print the table. You will use this information to answer questions on the activity worksheet.

10. If necessary you can also select the Chlamydia Tables, Gonorrhea Tables and the Syphilis Tables for additional statistical information to assist you in answering the questions on the activity worksheet.

ACTIVITY 2

1. Using Netscape or another Web browser, type the following address at the URL prompt:

   http://sunsite.unc.edu/ASHA/

2. This site, maintained by the American Social Health Association, is a user friendly guide to sexually transmitted disease. Use this homepage to find additional information on the diseases you were evaluating statistically in activity one. This site is also a good resource with the answers to the most frequently asked questions about sexually transmitted disease.

SEXUALLY TRANSMITTED DISEASE ACTIVITY WORKSHEET

Name _____   Date _____

Class/Section _____

1. How do you account for the differences in reported cases of gonorrhea between men and women?

2. As related to question one, why would you expect approximately equal numbers of cases of syphilis in men and women (as shown in table 2)?

3. What is the ratio of total *Chlamydia trachomatis* infection in men versus women (1995)? How would you account for this dramatic difference?

PART SEVEN

ECOLOGY AND BEHAVIOR

47

HUMAN POPULATION TRENDS

INTRODUCTION
Populations are groups of a single species occupying a particular geographical area. We frequently discuss the population of the United States, refer to a population of students, or describe the population of mosquitoes in the swamp. Very large populations are often thought of as being detrimental. Large populations of rodents can infest a neighborhood, large populations of locusts can decimate a crop, and large populations of human create tons of waste. However, small populations of animals are very susceptible to environmental change and are frequently the victims of extinction. Obviously, populations that are neither too large nor too small are just right.

In nature, appropriate population sizes are maintained by various limiting factors in the ecosystem. These limiting factors help establish the carrying capacity of the ecosystem. Have or will humans exceed the carrying capacity of the planet? That point is the subject of debate. Do you know what the population of the planet is? Even if we don't know exact numbers, we do know and can demonstrate that large human populations are the cause of increased air and water pollution, land degradation, waste accumulation, and depletion of natural resources.

By studying population distribution, demographics, and trends, it is possible to predict future populations and prepare for future environmental reaction. Perhaps we can discover ways to manage large human populations without destroying our planet.

CHECK FOR URL UPDATES
Before beginning the activities in this chapter, check the Wadsworth Biology Resource Center home page for URL updates and information:

http://www.wadsworth.com/biology

Once you have arrived at the Wadsworth Biology Resource Center home page, select General Biology and locate this book. Any updates should appear.

ACTIVITY

The U.S. Census Bureau has statistics for just about every population parameter you could imagine. This information is available on the U.S. Census Bureau's home page. Information on the size, distribution, and ethnic backgrounds of the population are provided. In this activity you will access the census home page and gather some data that you can use to dazzle and amuse your friends and family (not to mention your course instructor).

PART 1

1. Using Netscape, or another Web browser, type in the follow address at the URL prompt:

    http://www.census.gov/population/www

2. Select World.

3. Answer the questions that pertain to the world's population on the World and U.S. Population Worksheet.

4. Using the "Back" key of your Web browser, return to the previous page.

5. Select United States.

6. Answer the questions that pertain to the U.S. population on the World and U.S. Population Worksheet.

7. Using the "Back" key of your Web browser, return to the page with the heading "Population Topics."

PART 2

1. Select Population Estimates.

2. Select National Population Estimates.

3. Select Yearly Population Estimates by Age Group and Sex.

4. Select Population Data.

5. Answer the questions that pertain to specific groups of the U.S. population on the World and U.S. Population Worksheet.

Follow the directions from your instructor regarding the submission of your worksheets. Some instructors will want them e-mailed; others will want them handed in.

WORLD AND U.S. POPULATION WORKSHEET

Name _____   Date _____

Class/Section _____

PART 1

1. What is the population of the world at this minute on this date?

2. What is the projected population in two months?

3. How many births are occurring per hour?

4. How many deaths are occurring per hour?

5. What is the net gain in the number of humans per second?

6. What is the population of the United States on this day at this time?

PART 2

1. Since 1991, has the median age of the U.S. population gone up or down?

2. What is the largest five-year age group in the United States in 1996?

3. As of February 1, 1996, were there more males or females in the United States?

4. Consider the answers to the three previous questions, and try to extrapolate and predict the level of future increases in the human population in the United States.

48

PREDATOR-PREY INTERACTIONS

INTRODUCTION

The health of an ecosystem relies on the stability of food chains and food webs operating within that ecosystem. Predator-prey relationships in particular are very important at controlling populations of animals such that they remain within the carrying capacity of an ecosystem. If prey populations become very large, their numbers will eventually be reduced as they succumb to density dependent reduction factors. Reduction of the prey population will in turn make less food available to the predator ultimately resulting in a reduction of the predator population. Predator-prey relationships provide an excellent example of the interdependence of trophic levels in food chains.

In this activity you will model the interactions of predator-prey relationships by playing a game. In the game you will collect data on the population of your predator, the lynx, and its prey, the rabbit. After accumulating data, you will graph the population numbers and attempt to predict future population growth using your results and graph as model.

CHECK FOR URL UPDATES

Before beginning this activity, check on the Wadsworth Biology Resource Center homepage for URL updates and information:

http://www.wadsworth.com/biology

Once you have arrived at the Wadsworth Biology Resource Center homepage, select General Biology and locate this book. Any updates should appear.

ACTIVITY

1. Using Netscape or another Web browser, type the following address at the URL prompt:

    http://outcast.gene.com/ae/AE/AEPC/WWC/1991/predator.html

2. Read the entire experiment. Print the table provided in the experiment to record your data.

3. Carry out the procedure and record your data on the table provided. Prepare a graph of your populations, analyze results, and predict future populations.

49

CRITICAL ECOREGIONS

INTRODUCTION
Alteration or destruction of an ecosystem dramatically disrupts food chains and webs. Elimination of a food source, disruption of predator-prey relationships, and loss of habitat effectively unravels the food web resulting in collapse of the ecosystem.

The Sierra Club has identified critical Ecoregions throughout North America. This group identifies the Ecoregion's geographical location and develops a strategy to restore and maintain fully functioning ecosystems. The ultimate objective of this plan is to reestablish the "web of life" on earth.

In this activity you will identify and investigate the ecoregion closest to your home.

CHECK FOR URL UPDATES
Before beginning this activity, check on the Wadsworth Biology Resource Center homepage for URL updates and information:

http://www.wadsworth.com/biology

Once you have arrived at the Wadsworth Biology Resource Center homepage, select General Biology and locate this book. Any updates should appear.

ACTIVITY

1. Using Netscape or another Web browser, type in the following address at the URL prompt:

   http://www.sierraclub.org/ecoregions/

2. Select <u>Clickable Map of Critical Ecoregions</u>

3. Select the region on the map closest to your home.

4. Read and take note of the description and key objectives for your region

5. Answer the questions on the activity worksheet as they relate to your ecoregion.

CRITICAL ECOREGIONS ACTIVITY WORKSHEET

Name _____   Date _____

Class/Section _____

1. Name and describe the general characteristics of your ecoregion.

2. Identify the threats against the health of your ecoregion.

3. What are the objectives for repair and restoration of your ecoregion?

4. What activities, if any, can you become involved in to assist in implementation of the Sierra Club objectives?

50

THE OCEAN PLANET

INTRODUCTION

Earth, the water planet. Over 70 percent of the Earth's surface is covered by water. The oceans are a vast and largely unexplored ecosystem. Unlike terrestrial ecosystems, marine ecosystems are relatively uniform. However, two limiting factors—sunlight and dissolved oxygen—play a major role in establishing two major marine habitats with very different levels of productivity. The two habitats are the neritic zone and the oceanic zone. The neritic zone accounts for only 10 percent of the oceans surface but is responsible for over 80 percent of the world's commercial fishing harvest. Marine life in this zone is very abundant and diverse. This productivity is due to an abundance of mineral nutrients and sunlight. The oceanic zone that covers the deep water that stretches beyond the continental shelf has a minimal availability of nutrients and a lack of sunlight. As a consequence, this ecosystem is simple and not as diverse as the one located in the neritic zone.

Until recently, the vast open waters of the oceans had remained relatively unexplored (or exploited); as a consequence, the open seas have sustained minimal damage due to human activity. Unfortunately, that situation is changing. As we search for alternative food sources, and for minerals and petroleum on the continental shelf, we affect the health of marine ecosystems. It is important to maintain healthy aquatic ecosystems not only because of our quest for biological diversity, but also because terrestrial and marine ecosystems are intimately linked. Any change in one ecosystem is likely to affect the other. As animals of terrestrial ecosystems, it is therefore in our own best interests to protect these marine environments.

CHECK FOR URL UPDATES

Before beginning the activities in this chapter, check the Wadsworth Biology Resource Center home page for URL updates and information:

http://www.wadsworth.com/biology

Once you have arrived at the Wadsworth Biology Resource Center home page, select General Biology and locate this book. Any updates should appear.

ACTIVITY

In April 1995 the Smithsonian opened the Ocean Planet traveling exhibit. The objective of this display is to disseminate knowledge of the Earth's marine ecosystems, thereby promoting the importance of these ecosystems. It is hoped this knowledge will lead to greater understanding and conservation of the world's oceans. Although we will only travel through a small portion of this exhibit you are invited to return to the site when you have time available to explore this vast, electronic exhibition.

In this activity you will take a guided tour of the Ocean Planet exhibit. The subject of the tour will be pollution. The subjects covered on the tour will range from oil pollution to toxic contaminants to sewage to alien species. At the end of the tour you'll get an opportunity to post a note and express your own opinion concerning the health of the oceans. In addition, you may suggest possible solutions to the problem of marine pollution.

1. Using Netscape, or another Web browser, type in the following address at the URL prompt:

   http://seawifs.gsfc.nasa.gov/ocean_planet.html

2. Read the introductory page and then select Enter Exhibition Here.
3. Select Special Tour.
4. Select Pollution.
5. Under the page's list of your tour you can see that you will have twenty-one pages to visit. If you are short of time, you may choose to omit the first four pages by beginning with page 5. To being the tour here select Smart SEA Shopping.
6. Continue your tour through the remaining pages by selecting the "Next" button at the top of the page.
7. Once on page 21, you may see several messages that other visitors have left about the exhibit or ocean conservation. You may also read about some of the things you can do to preserve and protect our oceans. And finally, you can leave a message for others. You may comment on the exhibit or suggest ways the oceans can be maintained in a healthy state.
8. Once you have ended your tour, complete the questions found on the Ocean Planet Activity Worksheet located at the end of this chapter.
9. To further explore this site, you may select another guided tour or enter any room of the exhibit through the map of the Ocean Planet exhibition. You can return to the map by using the "Back" key of your browser.

Follow the directions from your instructor regarding the submission of your worksheets. Some instructors will want them e-mailed; others will want them handed in.

OCEAN PLANET ACTIVITY WORKSHEET

Name _____ Date _____

Class/Section _____

1. What is the largest source of oil contamination in the ocean?

2. Describe the possible sources of each of the following toxic ocean contaminants:

    Mercury

    Dioxin

    PCBs

3. What is the "Law of the Sea," and how does it control the dumping of marine debris?

The Ocean Planet

4. List three alien species of marine ecosystems as well as the sources of these species.

5. What are the populations of the four largest coastal cities?

6. How do these large coastal communities affect coastal waters?

51

TROPICAL RAINFORESTS: HERE TODAY, GONE TOMORROW

INTRODUCTION

There are six major terrestrial biomes spread over the surface of the planet. Each biome is named or characterized by the type of plant life that predominates the landscape. The plant life, of course, is dictated by the climate (rainfall and temperature) in the area. Thus, the grassland biome is characterized by a lack of trees, a predominance of grasses, and an annual rainfall of between 10 and 30 inches with cold winters and warm summers.

By far the most diverse and complex terrestrial biome is the tropical rainforest. This rich biome is thought to contain over 50 percent of the world species, including untold numbers of insect species, most of which have not yet been described. Rainfall in the tropical rainforest is abundant with over 80 inches of rain per year. Temperatures remain warm allowing a year-round growing season. Not only is the rainforest important as a home to many of the planet's species, but this massive concentration of plant life also serves as the major carbon dioxide sink on Earth. When tropical rainforests are cut down and then burned, we are yielding a double-edged sword. Not only do we lose a reservoir for storing carbon dioxide, but also by burning plants that store carbon dioxide, we are releasing stored CO_2 into the atmosphere. Unfortunately, tropical rainforest acreage is disappearing at an alarming rate. Development, agriculture, harvesting of tropical woods, oil exploration, and simple mismanagement have lead to the destruction of approximately 300 million acres of rainforest per year. At the current rate of destruction, tropical rainforests will be seriously depleted in our lifetime. This destruction may actually alter rainfall patterns in the temperate regions of the world. This could have a devastating effect because the temperate regions of the planet produce most of the food that feeds the world.

Fortunately, many environmental groups have taken up the cause of preserving and protecting the tropical rainforest biome. One very important and effect advocate is the Rainforest Action Network. Through this group's efforts, political, economic, and social pressure has been placed on many industries and governments involved in rainforest destruction.

CHECK FOR URL UPDATES

Before beginning the activities in this chapter, check the Wadsworth Biology Resource Center home page for URL updates and information:

http://www.wadsworth.com/biology

Once you have arrived at the Wadsworth Biology Resource Center home page, select General Biology and locate this book. Any updates should appear.

ACTIVITY

In this activity you will access the home page of the Rainforest Action Network (RAN). While at this site you will read about the rainforest, take a quiz to see how much you've learned about the rainforest biome, and sample some information regarding RAN campaigns and victories. While you're reading, remember that the information you are receiving at this site is the opinion of RAN. If any of the companies or government institutions mentioned at this site had the opportunity to present their opinions on their activities they would express a very different point of view.

PART 1

1. Using Netscape, or another Web browser, type in the following address at the URL prompt:

   http://www.ran.org/ran/

2. Select the Rainforest Information box.

3. Begin the activity by reading the information provided in the "Rainforest Information" section of the site.

4. Select Why Rainforests Are Important! Read the text.

5. Select Rates of Rainforest Destruction. Read the text.

6. Select Take a Rainforest Quiz. After answering a question, wait until the answer comes up on the screen with additional information before proceeding to the next question.

7. After completing the quiz, return to the home page by selecting the "Home page" box at the bottom of the page.

PART 2

1. Select Campaigns.

2. Select the World Bank. Read the text.

3. Answer the questions concerning the World Bank in the Tropical Rainforest–World Bank Activity Worksheet.

4. Return to the home page by selecting the "Home page" box at the bottom of the page.

PART 3

1. Select <u>Victories</u>.

2. Read about the concessions by industry and government institutions as a direct response to RAN-sponsored activities.

3. Answer some questions concerning these victories on the Tropical Rainforest–Victories Activity Worksheet.

Follow the directions from your instructor regarding the submission of your worksheets. Some instructors will want them e-mailed; others will want them handed in.

TROPICAL RAINFOREST—WORLD BANK ACTIVITY WORKSHEET

Name _____ Date _____

Class/Section _____

1. What is the mission of the World Bank?

2. Describe the four major funding project areas of the World Bank.

3. How has the World Bank contributed to or promoted ecologically destructive projects?

4. Describe two specific World Bank projects that resulted in environmental harm.

TROPICAL RAINFOREST—VICTORIES ACTIVITY WORKSHEET

Name _____ Date _____

Class/Section _____

1. Describe or discuss three different areas or organizations that the RAN has been able to influence through its activities.

2. Do you consider that any of these "victories" have played a significant role in the ultimate preservation of the tropical rainforest?

52

ENVIRONMENTAL TECHNICAL INFORMATION PROJECT

INTRODUCTION

We live in a global community. We can easily travel from one continent to another in less than one day. We can call a friend in Europe or Asia in less than one minute. Most of our clothing, electronic equipment, vehicles, and toys come from manufacturers in other countries. Unfortunately, we also share the consequences of pollution events with other countries. Activities occurring in one part of the world can have a detrimental effect on ecosystems found in other parts of the world. An example of traveling pollution close to home is the formation of acid precipitation in Ohio, which then affects sensitive pine forests to the north in Canada or to the south in North Carolina. Use of banned pesticides in African countries, release of chlorofluorocarbons in China, and nuclear testing by France are all issues of concern for every citizen of the world. Therefore, it is important that agreements on the maintenance, and conservation of the environment be a primary goal of all countries. Treaties or pieces of legislation can assure the commitment of both large and small countries to a common cause: a clean and healthy world.

CHECK FOR URL UPDATES

Before beginning the activities in this chapter, check the Wadsworth Biology Resource Center home page for URL updates and information:

http://www.wadsworth.com/biology

Once you have arrived at the Wadsworth Biology Resource Center home page, select General Biology and locate this book. Any updates should appear.

ACTIVITY

In this activity you will access the ECOLOGIA Web site. ECOLOGIA stands for ECOlogists Linked for Organizing Grassroots InitiAtives. This international group is headquartered in Pennsylvania and is intended to provide training and technical support for grassroots organizations in the former Soviet Union. ECOLOGIA is jointly sponsored by The National Institute for Environmental Renewal, The United States Agency for International Development, and the W. Alton Jones Foundation.

The objective of the site is to select the best environmental information sources and link them together at one location. ECOLOGIA's database topics include toxic chemicals, radiation issues, waste management, environmental legislation and treaties, energy-related technologies, global issues, and environmental education. In this activity you will become familiar with two pieces of legislation covering world wide conservation and pollution concerns.

PART 1

1. Using Netscape, or another Web browser, type in the following address at the URL prompt:

   http://ecologia.nier.org/index.html

2. Select <u>International Treaties and Global Issues</u>.
3. Find "The United Nations Framework Convention on Environmental Change." Under that heading select the <u>Beginners Guide to the Climate Change Convention</u>.
4. Read the text of the article (you may choose to print the text) and then answer the questions on the ECOLOGIA Activity Worksheet.

PART 2

1. Using the "Back" key of your browser return to the page listing all environmental legislation and treaties.
2. Find the heading "The Convention on Biological Diversity." Under that heading select <u>Convention on Biological Diversity (full text)</u>.
3. Read through the text (you may choose to print the text).
4. Answer the questions on the ECOLOGIA Activity Worksheet relative to this piece of legislation.
5. Using the "Back" key of your browser return to the page listing all environmental legislation and treaties.
6. Find the heading "The Convention on Biological Diversity." Under that heading select <u>Ratification Update</u>.
7. Read through the list of countries, making a mental note of any large, developed countries that appear not to have ratified the treaty.
8. Answer the question regarding ratification on the ECOLOGIA Activity Worksheet.

Follow the directions from your instructor regarding the submission of your worksheets. Some instructors will want them e-mailed; others will want them handed in.

ECOLOGIA ACTIVITY WORKSHEET

Name _____ Date _____

Class/Section _____

PART 1

1. What environmental issue is addressed in this agreement?

2. How is the evolution of humans tied to climate trends?

3. What is the ultimate objective of this treaty?

4. What group of countries will be responsible for technology transfer and financing of this treaty? Do you think this burden is fair?

Environmental Technical Information Project

PART 2

1. What is the objective of this treaty?

2. What countries will bear the financial burden of this treaty?

3. What large developed country(ies) is (are) missing from the ratification list? How can you explain this omission? Can you come up with a particular reason why a country would not consider ratifying this convention document?

53

BUTTERFLY AND BIRD MIGRATION

INTRODUCTION
Behavior is defined as a reaction to stimuli that can either be instinctive or learned. One very important behavior is migration. Migration is movement between regions that may be prompted by a search for food, warmer climes, or a preferred mating area. Migration is considered a learned behavior, whereas the ability to navigate the migration route is often innate. By studying both historic and present-day migration patterns, it is possible to evaluate the health of an ecosystem or species. For example, migrating shore birds rely heavily on native wetlands to support them as they travel many miles during migration. Destruction of wetland stopovers means loss of food along the way to the final destination. Sightings of particular species of shorebirds along historic migration routes have decreased dramatically. This may indicate a drop in the population due to a loss of wetland feeding areas.

We frequently think of birds when we think of migration, but species as diverse as bats, turtles, whales, butterflies, and many herd animals also migrate long distances during very specific times of the year.

CHECK FOR URL UPDATES
Before beginning the activities in this chapter, check the Wadworth Biology Resource Center home page for URL updates and information:

http://www.wadsworth.com/biology

Once you have arrived at the Wadworth Biology Resource Center home page, select General Biology and locate this book. Any updates should appear.

ACTIVITY

In this activity you will access The Why Files. This site is a project of the National Institute for Science Education and is funded by the National Science Foundation. The Why Files promotes the association of science and major news events in an attempt to provide students and teachers with an explanation and discussion of current events in the context of science. The Why Files page opens with current news stories, but also has a Filed Why Files section with previous news stories. Filed files can be searched by key word(s). You will begin the activity by investigating The Miracle of Migration. Subtopics of migration include Monarch Butterfly Activity, Navigation Information, Flight Strategies, and Forest Fragmentation and Songbird Decline.

PART 1

1. Using Netscape, or another Web browser, type in the following address at the URL prompt:

    http://whyfiles.news.wisc.edu/index.html

2. Select Filed Why Files.

3. Select Search.

4. Enter "The Miracle of Migration" in the search box.

5. Select "more" at the bottom of the page.

6. Read about the recent rare snowfall that killed monarchs overwintering in Mexico.

7. After reading each butterfly topic, continue to select "more" at the bottom of the page until you finish with the monarch butterfly presentation. (The command "more" is not available on the last butterfly page.)

8. Answer the questions about the monarch butterfly on the Monarch Migration Activity Worksheet located at the end of this chapter.

PART 2

1. Return to the Chart Your Own Route page by repeatedly selecting the "Back" key of your Web browser or by selecting the caterpillar at the bottom of any page.

2. Select Bird Migration.

3. Continue reading all pages covering bird migration until you come to the last page, which doesn't have a "more" command.

4. Answer the questions regarding bird migration on the Bird Migration Activity Worksheet at the end of this chapter.

PART 3

1. Return to the Chart Your Own Route page by repeatedly selecting the "Back" key of your Web browser or by selecting the caterpillar at the bottom of any page.

2. Select <u>Songbird Survival</u>.

3. Read the material covering songbirds, tropical deforestation, and forest fragmentation.

4. Answer the question on the Songbird Activity Worksheet at the end of this chapter.

PART 4, OPTIONAL

1. Return to the home page of this site by selecting the chrysalis at the bottom of the page.

2. Visit the Why Files Forum by selecting the FORUM icon.

3. In the Why Files Forum, you can post a note on several current topics. If you prefer you can take part in the general discussion on any topic.

4. As an optional activity, please take part in the online discussion of a current topic or a topic of your choice.

5. Monitor the discussion for a few weeks to see if anyone responds to your comments.

6. Summarize your discussion forum experience (your comments as well as the responses of others) on the Why Files Forum Activity Worksheet at the end of this chapter.

Follow the directions from your instructor regarding the submission of your worksheets. Some instructors will want them e-mailed; others will want them handed in.

MONARCH MIGRATION ACTIVITY WORKSHEET

Name _____   Date _____

Class/Section _____

1. What was the actual kill rate of monarchs in Mexico in December 1995 due to snowfall?

2. Describe the difference in migration behavior of the reproductive butterflies versus the migration butterflies.

3. Why do monarchs migrate specifically to Angangueo, Mexico?

4. How are monarch habitats being threatened both in Mexico and in summer reserves?

BIRD MIGRATION ACTIVITY WORKSHEET

Name _____ Date _____

Class/Section _____

1. Describe four ways in which birds navigate migration routes.

2. Flapping/gliding and bounding are two flight strategies birds employ when migrating. Describe each of these physical strategies, and give an example of a particular bird that you have observed employing each strategy. (If you can't think of an example, post a question on the Why Files Forum.)

3. List the mileage each of the following species has been known to migrate in one year:

   Arctic tern

   Whooping crane

   Barn swallow

4. List two historic influences on migration patterns.

5. Why are raptors restricted to flights over the Mexican Peninsula as they fly between North and South America?

SONGBIRD ACTIVITY WORKSHEET

Name _____ Date _____

Class/Section _____

How has forest fragmentation and cowbird parasitism contributed to the decline in songbird populations?

WHY FILES FORUM ACTIVITY WORKSHEET

Name _____ Date _____

Class/Section _____

1. Type your posted forum message below.

2. List and discuss any comments posted on the forum in response to your message.

Butterfly and Bird Migration

54

SEARCHING THE WEB ON YOUR OWN

INTRODUCTION

Up to this time you have been led around the World Wide Web through various sites and hyperlinks. The destinations and directions have been designed to limit the amount of time you spend searching for information. But ultimately everyone gets the urge to utilize the Web's search engines to find a particular topic in cyberspace. As with all things on the Web, the number of search engines continues to grow daily. Infoseek Guide, Lycos, Magellan, Excite, and Yahoo are but a few of the popular search services available.

After selecting the "Net Search" key on your browser, selecting your search engine, and then entering a key word or words, you will be presented with a list of Web sites that correspond to your search word(s). The best match will be listed first with less perfect matches listed in order of suitability. Remember, just because your key word has been found doesn't mean the site will yield much, or even appropriate, information. Be cautious as you begin your search. Before you know it, hours may have passed by and you may have discovered little information of use. Then again, if you stumble on a five-star site, you'll be rewarded for your efforts.

CHECK FOR URL UPDATES

Before beginning the activities in this chapter, check the Wadsworth Biology Resource Center home page for URL updates and information:

http://www.wadsworth.com/biology

Once you have arrived at the Wadsworth Biology Resource Center home page, select General Biology and locate this book. Any updates should appear.

ACTIVITY

In this activity you will choose a subject of particular interest to you. It, of course, should have something to do with biology. Select one of the Web sites presented to you following your query, and then evaluate the site for reliability and appropriateness of information.

PART 1

1. Using your Web browser's tool bar, select the "Search" or "Net Search" key.

2. Select your search engine by selecting one of the engines listed, or type in the name of the engine that you would like to use.

3. Enter your key word or words. (For example, look for information on the "Arctic Tundra Biome.")

4. Use your enter key or select the "Search" box, to enter the query.

5. Select one of any of the addresses that are presented by the search engine by clicking on the site address.

6. You can shop around the rest of the sites later, but for now, evaluate and rate the site for reliability, and presentation. Is the site presenting the material the way you expected it would?

7. If you have time and your browser supports "What's New" and "What's Cool" keys, use those keys to find interesting sites on various biological subjects.

8. Record your search parameters and results on the Search Activity Worksheet.

PART 2

It is possible to limit your search to the general topics of science or biology when using the search command. The following two URLs can be used to search only those sites related to biology.

The URL for "Excite" is:

http://www.excite.com/Subject/Science/s-index.h.html

To search Yahoo for biological sites, enter the following address:

http://gnn.yahoo.com/gnn/Science/Biology

1. Try using either of these URLs to search for the topic you entered in Part 1 of this activity.

2. Compare your results (time to locate, quantity, and quality of sites) from the random search and the "science"-directed search.

Follow the directions from your instructor regarding the submission of your worksheets. Some instructors will want them e-mailed; others will want them handed in.

SEARCH ACTIVITY WORKSHEET

Name _____ Date _____

Class/Section _____

1. Which search engine did you choose?

2. What key word(s) did you search for?

3. What is the address of the site you used following your search?

4. Did your site prove reliable or useful? Why or why not?

5. Did you find any biologically "cool" sites?